Power Electronic Converters Modeling and Control

Power Electronic Converters Modeling and Control

Editor

Beniamino Cipriani

Power Electronic Converters Modeling and Control

Edited by **Beniamino Cipriani**

Printed in 2017

ISBN: 978-1-68117-166-1
Library of Congress Control Number: 2015951998

© 2016 by
SCITUS Academics LLC,
616, Corporate Way, Suite 2, 4766,
Valley Cottage, NY 10989

www.scitusacademics.com

Preface

Modern power electronic converters are involved in a very broad spectrum of applications: switched-mode power supplies, electrical-machine-motion-control, active power filters, distributed power generation, flexible AC transmission systems, renewable energy conversion systems and vehicular technology, among them.

Electronic power conversion has gained widespread acceptance in power systems applications; electronic power converters are increasingly employed for power conversion and conditioning, compensation, and active filtering. This book presents the fundamentals for analysis and control of a specific class of high-power electronic converters-the three-phase voltage-sourced converter. Voltage-sourced converters in power systems provides a necessary and unprecedented link between the principles of operation and the applications of voltage-sourced converters.

Power Electronics Converters Modeling and Control teaches the reader how to analyze and model the behavior of converters and so to improve their design and control. Dealing with a set of confirmed algorithms specifically developed for use with power converters.

Table of Contents

CHAPTER 5   **Modeling and Current Programmed Control of a Bidirectional Full Bridge DC-DC Converter ....**
Error! Bookmark not defined.17

CHAPTER 6   **PI and RST Control Design and Comparison for Matrix Converters Using Venturini Modulation Strategy ...**
Error! Bookmark not defined.47

CHAPTER 1

Experimental Assessment of Derating Guidelines Applied to Power Electronics Converters

S.E. De León-Aldaco[1], J.H. Calleja-Gjumlich[2], H.R. Jiménez-Grajales[3] and F.I. Chan-Puc[4]

[1,2] Departamento de Ingeniería Electrónica Centro Nacional de Investigación y Desarrollo Tecnológico Cuernavaca, Mor., México
[3] Gerencia de Energías no Convencionales Instituto de Investigaciones Eléctricas Cuernavaca, Mor., México
[4] División de Ciencias e Ingeniería Universidad de Quintana Roo Chetumal, Q. Roo., México

ABSTRACT

Power transistors are the most vulnerable components in switching converters, and derating is usually applied to increase their reliability. In this paper, the effectiveness of derating guidelines is experimentally assessed using a push-pull DC-DC converter as a case study, operating in three different environments. After measuring the electrical variables and temperature, reliability was predicted following the guidelines in MIL HDBK 217F. The sensitivity analysis performed indicates that temperature has the largest impact on reliability, followed by environment and device quality. The results obtained demonstrate that a derating procedure based solely on DC ratings does not ensure an adequate performance. Therefore, additional guidelines are suggested to help increase the overall reliability obtained from a power circuit.

INTRODUCTION

Reliability is usually defined as the ability of an item to perform a required function under stated conditions for a stated period of time [1]. It is widely recognized that any competitive industry must know the reliability of their products, has to be able to control it, and should produce at the optimum reliability level that yields the minimum life-cycle cost to the user [2]. However, this is not always the case. Recently, an industry-based survey aimed at determining the expectations and requirements of power electronics converters was conducted. The results confirm that reliability is indeed an area of concern and that better monitoring methods and indicators are needed [3].

The desired reliability level must be established at the design phase, because subsequent testing and production will not raise the reliability without a basic design change [4]. Thus, the first step is to select the configuration best suited to the task. It is not easy to identify the most reliable converter in a particular application, but several comparisons that might help for this purpose have been reported in recent years: comparison of three circuits aimed at grid-connected applications [5], of four rectifier-inverter topologies and one matrix converter [6], of a standard three-phase inverter and two redundant configurations [7], of two converters aimed at fuel-cell applications [8], and the comparison of two multilevel converters aimed at motor drive applications [9], among others. In any configuration, reliability depends heavily on the judicious selection of the individual components involved, and the most failure-prone ones must be carefully specified. Reliability is often expressed either in terms of the failure rate λ, or the mean time between failures MTBF. Table 1 lists the percent contributions of the components in several power converters to the overall failure rate.

Table 1. Percent failure rates for components in power electronics converters.

Ref	Application	Contribution to overall failure rate (%)			
		Diodes	Transistors	Capacitors	Magnetic elements
	Wound rotor induction generator	53.38	46.62	N.A.	N.A.
[10]	Permanent magnet generator	34.57	65.43	N.A.	N.A.
	Boost converter	2.91	75.7	20.85	0.54
	Forward converter	32.65	54.2	12.3	0.86
[11]	Single-phase PFC	12.19	76.52	10.17	1.12
	Boost converter, CCM, 1 kW	5.08	67.36	24.46	3.1
	Boost converter, DCM, 1 kW	3.75	76.74	17.42	2.09
	Boost converter, CCM, 300 W	4.89	67.43	24.14	3.54
[12]	Boost converter, DCM, 300 W	3.51	74.44	19.6	2.45
[13]	Automotive inverter	2.79	88	9.21	N.A.
	Average:	15.57	69.24	17.26	1.96

The components are broadly grouped in four categories: diodes, transistors (either MOSFETs or IGBTs), capacitors (regardless of the dielectric material), and magnetic elements, including inductors and transformers. According to Table 1, almost 69% of the failure rate can be attributed to the power transistors. In the opposite side, the contribution of the magnetic elements is quite small.

If reliability is to be improved, the weakest component should be upgraded first. A common approach is to derate the components (that is the intentional reduction of electrical, thermal and mechanical stresses on components to levels well below their specified rating) and several guidelines have been provided for critical applications [14]. It is usually assumed that derating provides lower failure rates, but the true effectiveness of the margins cannot be assessed unless the resulting reliability is calculated in the actual application.

This paper is aimed at experimentally assessing the effectiveness of derating the switching devices in a power circuit. A push-pull DC-DC converter rated at 100 W was selected for the analysis. Four prototypes were built according to a common design, but using transistors with different ratings. Electrical variables and circuit temperatures were measured at full load.

Afterwards, the resulting reliabilities were calculated following the procedure in the military handbook MIL HDBK 217F (MH217), taking into account three operating environments [15]. The sensitivity analysis indicates that temperature has the largest impact on reliability, followed by environment and device quality. The results obtained demonstrate that a derating procedure based solely on DC ratings does not ensure an adequate performance. After analyzing the experimental results, several guidelines are suggested to help increase the overall reliability obtained from a power circuit.

RELIABILITY PREDICTION

Let xi be the state of component i, such that $xi = 0$ if the component has failed, and $xi = 1$ if the component is functioning. Let XS be the state of a system comprised by n individual components. In a serial system (from a reliability point of view):

$$X_S = \prod_{i=1}^{n} x_i$$

(1)

That is, the system will fail if any of the components fails. In a parallel (or redundant) system:

$$X_S = 1 - \prod_{i=1}^{n} (1 - x_i)$$

(2)

Thus, the system remains functional when at least one of its components functions does too. Power electronics converters have become a commodity, and redundancy is seldom used. Competitive advantages are obtained from better performance, and longer operational lives [16][17]. Such is the case of most converters in PV application, which are of the series type.

The reliability of any type of item is a function of the item's failure rate λ which, for any electronic component, is assumed to be constant. The reliability $R(t)$ is then

$$R(t) = \exp(-\lambda t)$$

(3)

Photovoltaic converters are expected to operate continuously for as long as possible, and a specific lifetime does not exist. Therefore, mission reliability is not the best way to specify reliability requirements. A more useful parameter for continuously operating items is the mean time between failures MTBF, given by

$$MTBF = \int_{t=0}^{\infty} R(t)\,dt = \frac{1}{\lambda}$$
(4)

or, for a system comprised by n individual components:

$$MTBF = \frac{1}{\sum_{i=1}^{n} \lambda_i}$$
(5)

If the failure rate for a power converter is to be predicted using (5), then the individual failure rates for all the components in the converter must be computed first. Failure rates can be predicted from observed laboratory or field data, according to statistical analysis methods. One possible source of data is the MH217, which contains information for all types of components operating under prescribed conditions, obtained by collecting data since the late 50s. According to the model in MH217, the failure rate λ_P for any given part can be computed as:

$$\lambda_P = \lambda_b \prod \pi_j$$
(6)

where λ_b is the base failure rate under the prescribed conditions, and tabulated in the handbook. The π_j terms are stress factors that take into account the severity of the particular operational conditions. Those applying to the components usually employed in a power electronics converter are listed in Table 2. The temperature factor π_T is based on the Arrhenius equation as follows:

Table 2. Stress factors

Component	π_T	π_Q	π_E	π_A	π_C	π_V	π_S	π_P
MOSFET	•	•	•	•				
Resistor	•	•	•				•	•
Transformer	•	•	•					
Diode	•	•	•			•		
Capacitor	•	•	•		•	•		
Inductor	•	•	•					

$$\pi_T = exp\left\{\frac{-E_a}{K_B}\left(\frac{1}{T_X+273} - \frac{1}{298}\right)\right\}$$

(7)

The correspondences to TX and the activation energy Ea values are listed in Table 3. The term $kB = 8.617$ x 10^{-5} eV/°K corresponds to the Boltzmann's constant.

Table 3. Variables for the calculation of π_T.

Component	E_a(eV)	T_X(°C) (Temperature at)
Power Transistor	0.166	Junction T_J
Diode	0.266	Junction T_J
Capacitor	0.15	Ambient T_A
Inductor	0.11	Hot spot T_{HS}
Transformer	0.11	Hot spot THS
Snubber resistor	0.08	Component T_R

The junction temperature T_J, in °C, can be calculated using:

$$T_J = T_{CASE} + P_d\theta_{JC}$$

(8)

where T_{CASE} is the case temperature, Pd corresponds to the power dissipated at the device, and θ_{JC} is the thermal resistance from junction to case. The hot-spot temperature T_{HS} for magnetic components can be calculated using:

$$T_{HS} = T_A + 1.1\,\Delta T$$

(9)

where T_A is the ambient temperature. ΔT is the average temperature rise above TA, and can be approximated using:

$$\Delta T = \frac{125 P_d}{A}$$

(10)

where P_d corresponds to the power dissipated (W) and A is the radiating surface area of the component's case (in^2).

The environment stress factor π_E ranges from ground benign to cannon launching. The evaluations reported herein include ground benign G_B (non-mobile, temperature and humidity controlled environments), ground fixed G_F (moderately controlled environments with adequate cooling air), and ground mobile G_M (equipment installed on wheeled vehicles).

The quality factor π_Q depends on the package (plastic, JAN qualified, and so forth). The factor π_A is the application factor for transistors, set according to the type of use of the transistor in the circuit (mainly the power rating, P_r). The capacitance factor π_C is determined by

$$\pi_C = C^{0.09}$$

(11)

The voltage stress factor π_V for the capacitor, is defined by

$$\pi_V = \left(\frac{S}{0.6}\right)^5 + 1$$

(12)

where S is the ratio of operating voltage to rated voltage. The electrical stress factor π_S for diodes is defined by

$$\pi_S = 0.054 \text{ for } V_S < 0.3$$

(13)

$$\pi_S = V_S^{2.43} \text{ for } 0.3 < V_S \leq 1$$

(14)

Where V_S is the ratio between the reverse voltage applied to the diode, and the corresponding rating. In resistors, π_S is the power stress factor defined by

$$\pi_S = 0.71 e 1.1 S$$

(15)

where S is the ratio of operating power to rated power. The power factor π_P for resistors is determined by

$$\pi_P = (P_d)^{0.39} \tag{16}$$

where P_d corresponds to the power dissipated.

PUSH-PULL CONVERTER

A push-pull DC-DC converter aimed at photovoltaic (PV) applications was selected as the case study. It is shown in Figure 1. The circuit provides galvanic isolation, an important feature in applications such as those encountered in the development of PV systems. The input and output voltages considered are $V_{in} = 17.2$ V, and $V_o = 48$ V.

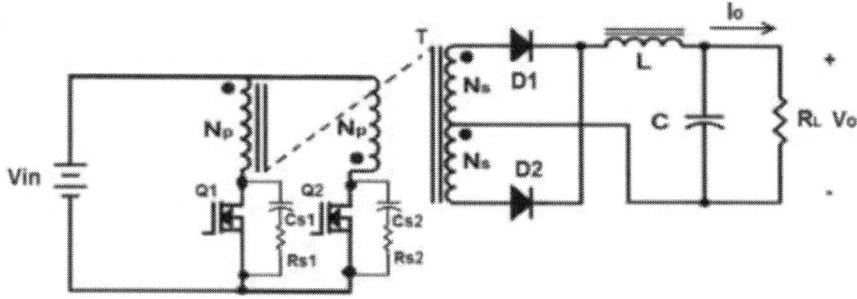

Figure 1. Push-Pull converter with RC snubber.

The following parameters were selected: switching frequency $fS = 100$ kHz, current ripple at the inductor $\Delta IL = 0.01\ Io$, and output voltage ripple $\Delta Vo = 0.05\ Vo$. According to the design procedure described in [18], the passive components requiredfor the output filter are L=600 µH, and C=47 nF. A simulation of the circuit yields the following results:

- Transistors: $I_{Q(AVE)} = 3.33$ A, $I_{Q(MAX)} = 10.91$ A, $V_{DS}=34.4$ V.
- Diodes: $I_{D(AVE)} = 1.042$ A, $I_{D(MAX)} = 2.183$ A, and $V_{RRM} =172$ V

The load ratio LR is defined as the magnitude of an electrical variable applied to a component, as a percent of the maximum rating. For MOS transistors, reference [15] recommends the following values: voltage V_{DS} load ratio = 80 %; current I_{DS} load ratio = 75 %. The ratings of the transistors used to build four prototypes are listed in Table 4. As can be seen, the recommended load ratios are not exceeded in the application.

Table 4. Main characteristics of switching devices.

Prototype	Device	V_{DSS}		I_D		$R_{DS(on)}$ (Ω)	θ_{JC} (°C/W)
		V	LR(%)	A	LR(%)		
P1	IRFP064	55	62.5	80	13.6	0.008	0.75
P2	IRFP044	55	62.5	37	29.5	0.02	1.3
P3	IRFZ40	50	68.8	32	34.1	0.028	1
P4	IRFP150	100	34.4	30	36.4	0.036	0.95

Two 600 V/15 A fast-recovery diodes were specified.

Both transistors operate in the hard-switching regime. Therefore, snubber networks were included in parallel with the switching devices in order to suppress the voltage spikes. The values selected for the snubbers are: C_{SN} = 22 nF, and R_{SN} = 18 Ω.

MEASUREMENTS

The tests were performed using as input a DC power supply instead of a PV module, thus providing equal operating conditions for the four prototypes. There were two sets of measurements performed. The first one involved the measurement of electrical variables in all the components. Figure 2 illustrates typical waveforms at the MOSFET transistors.

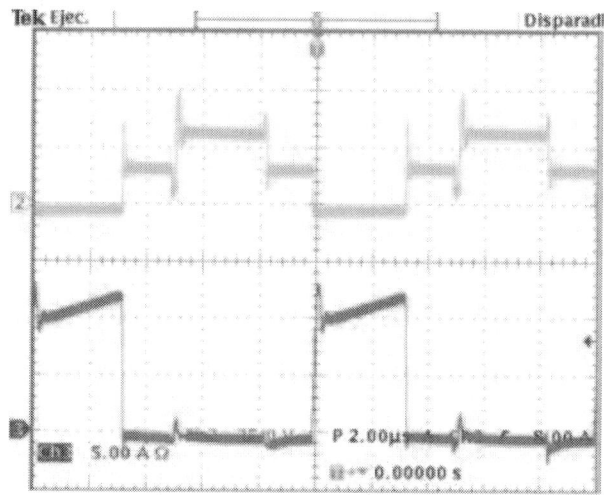

Figure 2. Waveforms at a transistor. Top trace: Drain to source voltage, V_{DS} (25 V/div). Bottom trace: Drain current, I_D (5 A/div). Time base: 2 µs/div.

The calculation of π_T requires the measurement of case temperatures in all the components comprising the converter. Temperatures were measured using a thermographic camera. Figure 3a shows the image of a prototype operating at full load. The transistor heatsinks are discernable at the top, left and right sides of the image; the hot spot at the bottom corresponds to the power transformer. Figure 3b illustrates the temperature of a diode mounted in a stamped heatsink.

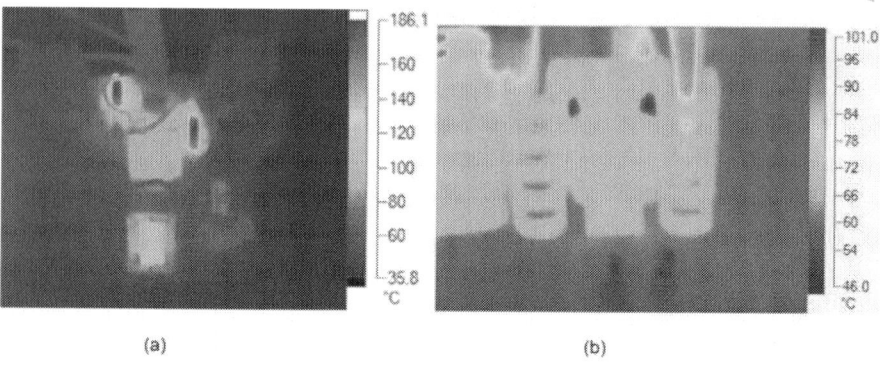

(a)   (b)

Figure 3. Thermographic images of the power converter. a) Top view. b) Detail of one of the diodes.

RESULTS

The stress factors for each component were calculated from maximum voltage and current data, power dissipation and temperature measurements. The procedure briefly described in Section II was implemented as mathematical routines in the Mathcad software package. The calculations were focused exclusively on devices in the power stage and do not include control circuits or drivers. A detailed example of the reliability calculations is included in Appendix A.

Figure 4 illustrates the failure rates for the components included in prototype P_1, operating in a ground fixed environment at an ambient temperature $T_A = 28$ °C. The percent contributions are illustrated in Figure 5. As expected, the most failure-prone devices are the power transistors, while the contribution of the inductor is negligible. Similar behaviors are exhibited by the other prototypes.

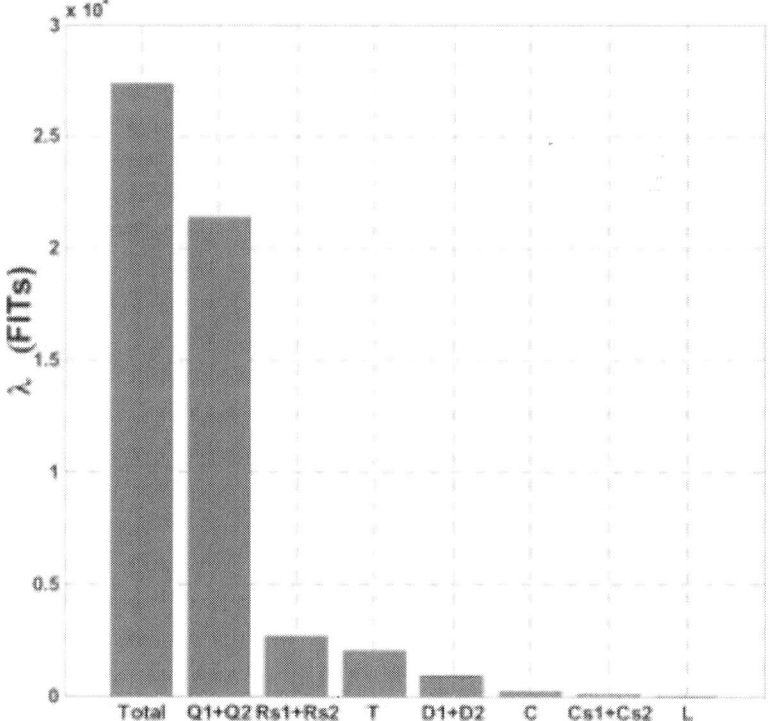

Figure 4. Failure rates for components in prototype P_1.

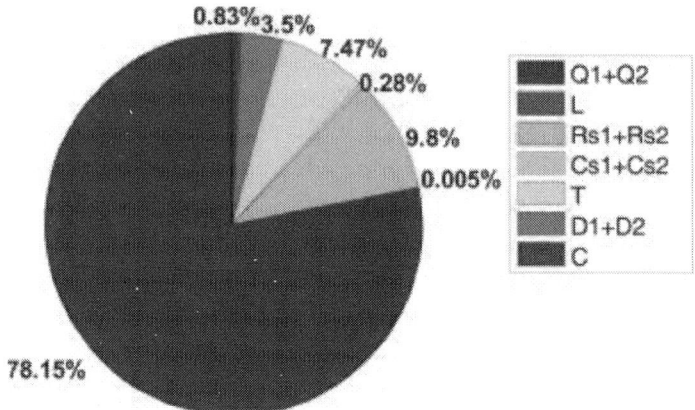

Figure 5. Percent contributions to the overall failure rate.

Figure 6 shows the contributions of each electronic component in the prototypes, evaluated in a ground fixed environment. It can be seen that, in the four converters, the largest contribution to the overall failure rate is associated with the MOSFET, and that the inductor produces the smallest contribution. The contributions of the capacitors are also small because they are of the metallized polypropylene type, which are less sensitive to temperature. The effect of the environment is illustrated in Figure 7.

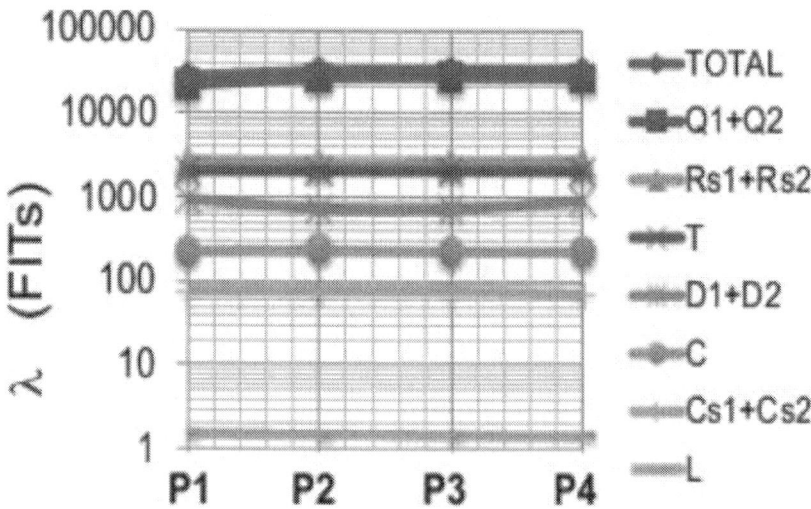

Figure 6. Contribution to the global failure rateof each electronic component.

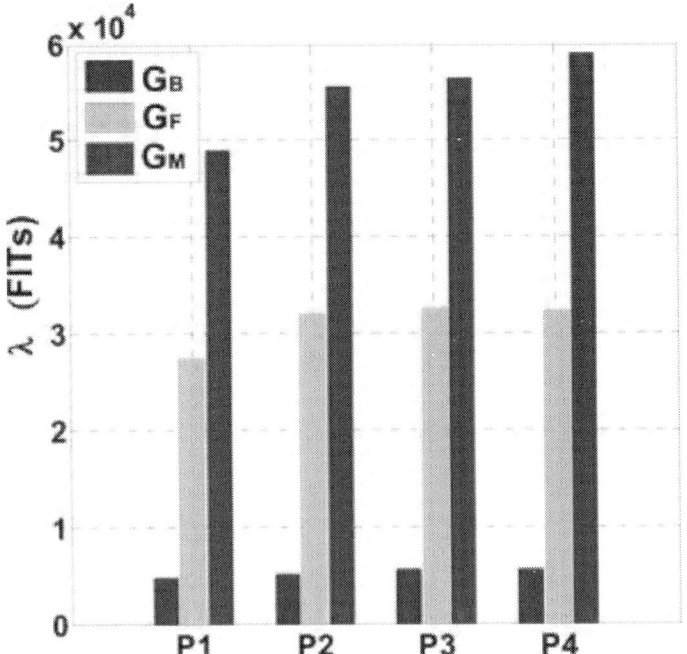

Figure 7. Overall failure rates as a functionof the operational environment.

DISCUSSION

Table 5 is a summary of the reliability obtained from each prototype in a ground fixed G_F environment, and the corresponding MTBF. The reliabilities as a function of time are plotted in Figure 8 (R_1 corresponds to prototype P_1, and so forth). The best performance is obtained from prototype P_1, built with the transistors that have the lowest on-resistance ($R_{DS(on)} = 0.008 \; \Omega$).

Table 5. Reliability.

Prototype	λ (FIT)	MTBF (years)
P_1	27397	4.16
P_2	32012	3.56
P_3	32545	3.5
P_4	32287	3.53

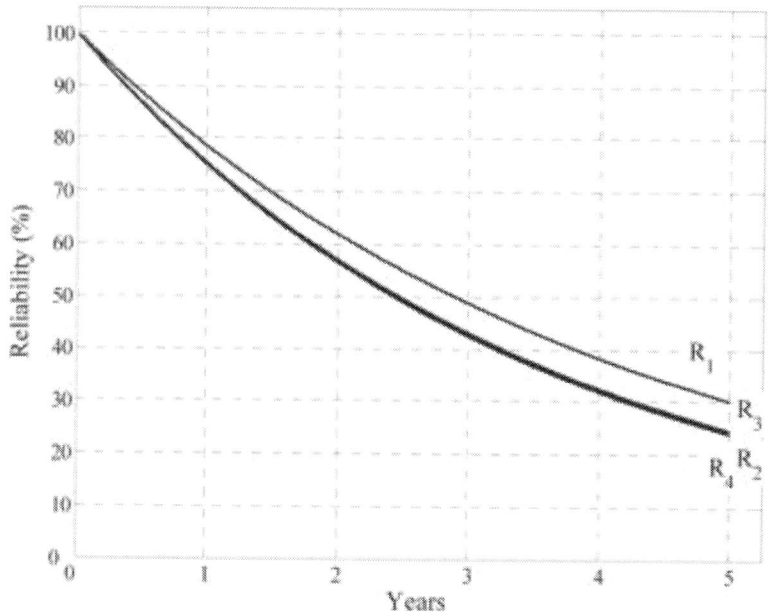

Figure 8. Reliability plots for the four prototypes.

Prototypes P_2, P_3 and P_4 exhibit similar reliabilities, without a clear relationship between this parameter and the on-resistances, or the load ratios. This aspect is illustrated in Figure 9 which depicts the relationship between the MOSFETs failure rates and on-resistances when the devices are mounted in an extruded heatsink with natural convection cooling in a ground fixed G_F environment. The coordinates ($R_{DS(on)}/\lambda$) are shown in the figure.

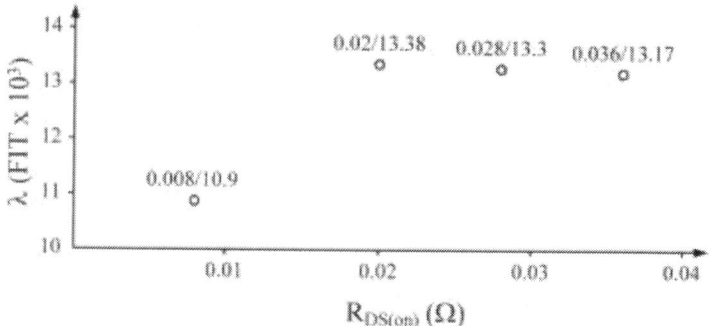

Figure 9. Failure rate as a function of the on-resistance.

A better understanding can be obtained by performing a sensitivity analysis; that is, analyzing the individual effects of the stress factors on the overall reliability of the converter. The analysis was carried out by varying one parameter within the limits allowed by the model used to predict the reliability, keeping the rest in its typical or nominal values. Prototype P_1 was used for temperature measurements, and the results are presented graphically in Figure 10. In the graph, the longest line corresponds to the parameter that has the greatest effect on λ. It can be readily appreciated that the reliability prediction model is highly sensitive to temperature. This behavior is corroborated by plotting the failure rates as a function of the transistors case temperatures, in Figure 11. It can be further appreciated that there are not noticeable differences among the prototypes.

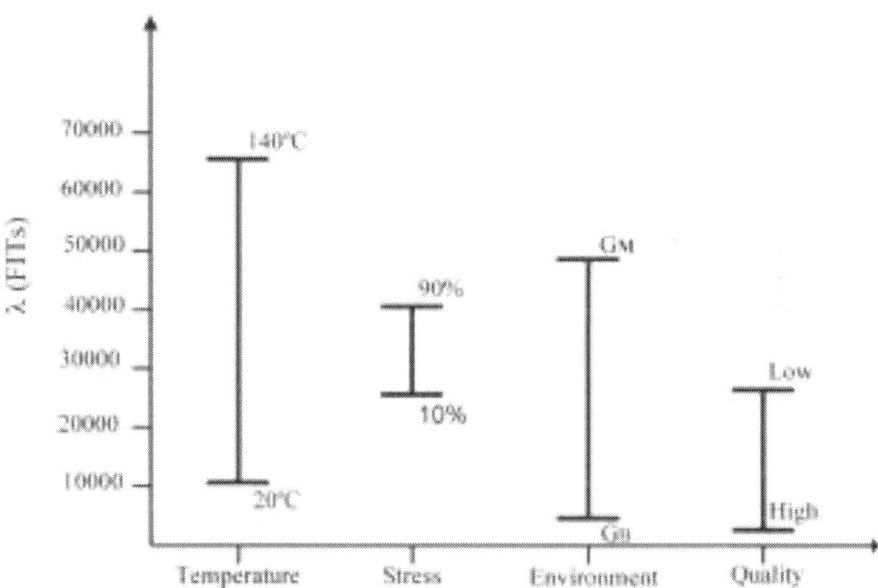

Figure 10. Results of the sensitivity analysis.

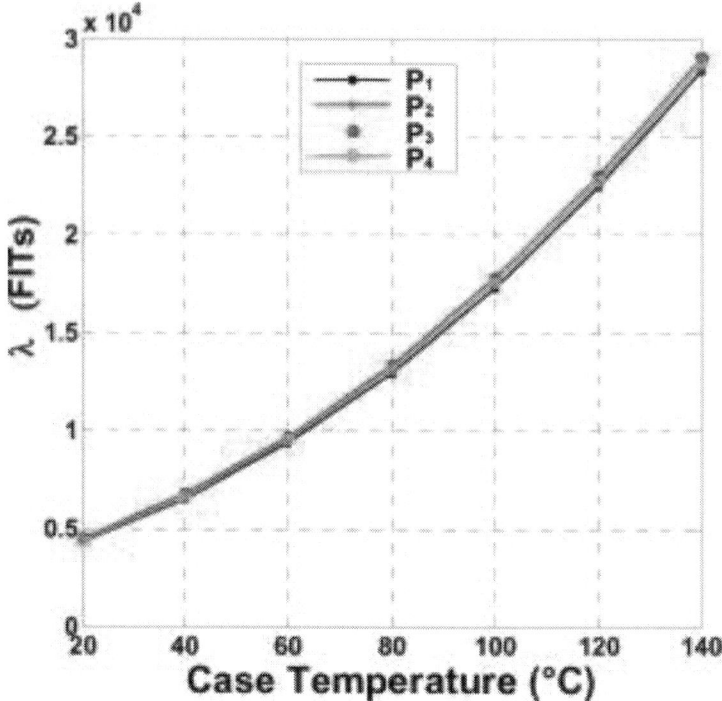

Figure 11. Failure rate over MOSFET case temperature.

According to the previous results, the components most prone to failure are the MOSFETs because they are highly sensitive to temperature. In order to calculate the junction temperature TJ using (8), $TCASE$ must be computed or measured first. The following equation applies:

$$T_{CASE} = T_A + P_d(\theta_{CS} + \theta_{SA}) \tag{17}$$

where θ_{CS} is the case-to-sink thermal resistance, which depends on the mounting technique although, for a given package, is fairly constant. The term θSA is the sink-to-ambient thermal resistance and depends basically on the surface area of the heatsink. Therefore, the overall failure rate depends, to a very large extend, on the size of the heatsinks. Figure 12 illustrates the dependency when both TA and Pd are constant. It is clear that, by choosing a low θ_{sa}, it is possible to obtain a more reliable converter.

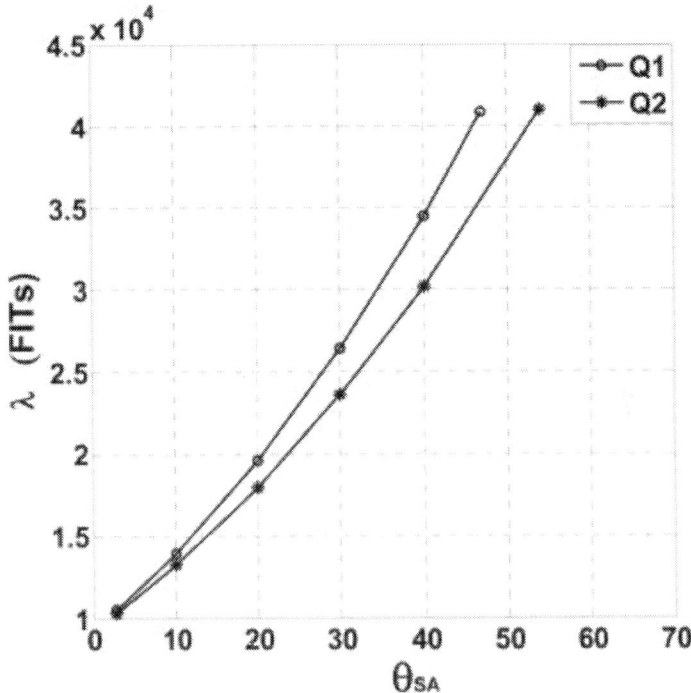

Figure 12. MOSFET failure rate *vs* θ_{SA}.

For further comparisons, two sets of tests were performed with prototype P_1, in each set the MOSFETs were attached to different heatsinks. The thermal resistance of the heatsinks used in the first set was θ_{SA}=2.7 °C/W. There were minor differences in the temperatures measured at the MOSFETs: T_{CASE}=69 °C for Q1, and T_{CASE}=67 °C for Q2. The resulting failure rates were λ_{Q1} =10901 FITs and λ_{Q2} =10510 FITs for Q1 and Q2 respectively. The thermal resistance of the heatsinks in the second set was θ_{SA}=4.4 °C/W, which produced higher case temperatures at both transistors. MOSFET Q_1 reached T_{CASE} = 82 °C and a corresponding λ_{Q1} = 13508 FITs. For Q2, T_{CASE} = 80 °C and λ_{Q2} = 13073 FITs.

Figure 13 shows thermographic images of one of the MOSFETs with each heatsink used. It is worth pointing out that both heatsinks have the same footprint, but different heights. The volume of the one used in the first set of tests is 66.68 cm³; the volume of the second one is 40.01 cm³. In the converter tested, reducing the hestsink volume by 40 % increases the failure rate by about 24 %.

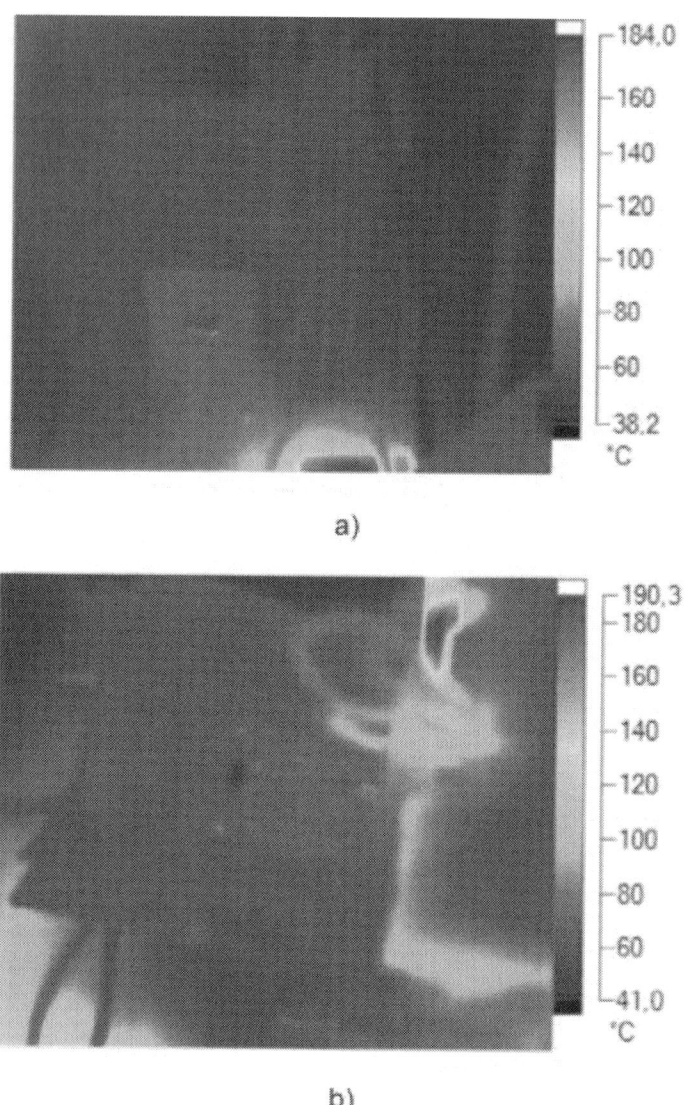

Figure 13. Thermographic images of the MOSFET. Detail view. (a) First set of tests, with θ_{SA}=2.7 °C/W. (b) Second set, with θ_{SA}=4.4 °C/W.

According to the previous results, it can be stated that:

- Temperature has indeed the largest effect on the failure rate. Thus, all available means should be employed to minimize the dissipation within the components. According to an analysis of losses in the transistors in prototype P_1, 73% corresponds to switching losses, 17% are conduction losses, and the remaining 10% are gate-related. The trends in the other prototypes are similar. These results clearly show that switching characteristics are far more important than the conducting behavior, and that a derating procedure based solely on the on-resistance or in the load ratios (as suggested in [15]) does not ensure the best performance.

- Quite often, soft-switching techniques are implemented in order to reduce the heating in the main transistors. It should be kept in mind, however, that in this case there are more components contributing to the overall failure rate. Care should be taken to ensure that the inclusion of the additional components does not offset the gains obtained from lower operating temperatures. Over-derating the transistors can be counterproductive because large devices might have larger gate capacitances which, in turn, will impose more exacting requirements on the drivers.

- Thermal management techniques are of paramount importance. Due to the exponential nature of (7), a small change in temperature produces a large variation in the failure rate. Within the limits set by volume and budgetary restrictions, the best heat sink available should be selected.

- According to Figure 11, electrical stresses have a rather limited impact on the failure rate. As stated above, using too large transistors does not necessarily provide the best reliability performance.

- There is no doubt that high quality components (as defined by MIL HDBK 217) achieve low failure rates. The drawback, however, is that including components of this kind usually results in much higher costs. Since there is currently a broad spectrum of low-cost plastic packages, a suitable approach is to specify the one with the best thermal performance (that is, the one with the lowest thermal resistances).

- The expected operating environment should be clearly identified. A converter that exhibits an acceptable performance in the ground-benign case will perform poorly when operated in a more demanding environment.

- Experimental results for magnetic elements reported herein indicate much larger failure rates than those reported in the references used to built Table 1 (almost four times the average value in Table 1). Thus, it seems that their failure rates are grossly underestimated. Reliability of these components is not to be taken as granted, and careful thermal performance evaluations must be carried out. The simplest way to increase the performance of magnetic components (from a reliability point of view) is to design them considering the lowest temperature rise allowed by volume and budgetary restrictions.

CONCLUSIONS

Throughout the years, it has been found that, in most power electronics converters, transistors contribute with the largest share to the overall failure rate. Quite often, heuristic derating guidelines are applied to deal with the vulnerabilities. In this paper, the experimental evaluation of derating the power transistors in a push-pull DC-DC converter is reported. The sensitivity analysis confirms that temperature has the largest impact on the failure rate, followed by environment and device quality. The results obtained demonstrate that a derating procedure based solely on the voltage and current ratings does not provide the best performance. The switching characteristics of the power transistors must be taken into account because they have a major impact on power dissipation. After analyzing the experimental results, additional guidelines are suggested to help increase the overall reliability obtained from a power circuit. Thermal management techniques and the expected operating environment are of paramount importance and must be included in any reliability improvement effort.

Appendix A

Table A1 lists measured and characteristic values for semiconductor devices. Table A2 includes similar information for capacitors and resistors, and Table A3 for magnetic elements. The results from calculations are listed in table A4. The terms λ_b, π_Q, π_E and π_A are obtained from [15]. The stress factors in the next six columns are obtained by applying the equations listed in the uppermost row. The overall failure rate λ_S is obtained by adding the contributions from the individual elements.

Table A1. Data and electrical measurements at semiconductors.

Component	$P_d(W)$	$T_{CASE}(°C)$	θ_{JC} (°C/W)	Vs
Q1	2.45	69	0.75	-
Q2	2.14	67	0.75	-
D1	3.7	92	1.3	0.29
D2	3.58	90	1.3	0.29

Table A2. Data and electrical measurements at passive components.

Component	T_A(°C)	T_{CASE}(°C)	S
C	28	-	0.811
C_{S1}-C_{S2}	28	-	0.811
R_{S1}-R_{S2}	-	225	0.8473

Table A3. Data and electrical measurements at magnetic components.

Component	Pd (W)	A (in²)	ΔT (°C)
T	2.2.	4.33	63.54
L	2.44	3.62	84.22

Table A4. Summary of reliability calculations for the Prototype P_1.

	λ_b(FIT)	π_Q	π_E	π_A	π_T(7)	π_C(11)	π_V(12)	π_S(13)(14)	π_S(15)	π_P(16)	λ_C(FIT)(6)
Q1	12	8	6	8	2.36	---	---	---	---	---	10901
Q2	12	8	6	8	2.28	---	---	---	---	---	10510.5
Rs1	3.7	10	4	---	3.5	---	---	---	1.8	1.44	1342.7
Rs2	3.7	10	4	---	3.5	---	---	---	1.8	1.44	1342.7
T	49	3	6	---	2.32	---	---	---	---	---	2047
D1	25	8	6	---	7.49	---	---	0.054	---	---	485.7
D2	25	8	6	---	7.13	---	---	0.054	---	---	462.36
C	0.51	10	10	---	1.06	0.76	5.51	---	---	---	226.4
Cs1	0.51	10	10	---	1.06	0.71	1	---	---	---	38.36
Cs2	0.51	10	10	---	1.06	0.71	1	---	---	---	38.36
L	0.03	3	6	---	2.83	---	---	---	---	---	1.53
										λS =	**27397 FIT**

REFERENCES

1. P. Kales. Reliability for Technology, Engineering and Management. Prentice-Hall, 1998. pp. 7-13.
2. O'Connor, P.. "Practical Reliability Engineering". John Wiley and Sons, 2002, pp. 414-452.
3. Yang S., Bryant A., Mawby P., Xiang D., Ran L., and Tavner P. An Industry-Based Survey of Reliability in Power Electronic Converters. IEEE Transactions on Industry Applications, Vol. 47, No. 3, May/June, 2011, pp. 1441-1451.
4. Ireson, W., Coomb, C., Moss, R. Handbook of reliability Engineering and Management. McGraw-Hill, 1996. pp. 5.1-5.16.
5. F. Chan F., Calleja H., and Martinez E. Grid connected PV systems: A reliability-based comparison. IEEE International Symposium on Industrial Electronics, 2006, pp. 1583-1588. Montreal, Canada, August.
6. Aten M., Towers G., Whitley C., Wheeler P., Clare J., and Bradley K. Reliability Comparison of Matrix and Other Converter Topologies. IEEE Transactions on Aerospace and Electronics Systems Vol. 42, No. 3, July 2006, pp. 867-875.
7. Julian A., and Oriti G. A Comparison of Redundant Inverter Topologies to Improve Voltage Source Inverter Reliability. IEEE Transactions on Industry Applications, Vol. 43, No. 5, September/October 2007, pp. 1371-1378.
8. Ranjbar A., Abdi B., Nabavi-Niaki S., Gharehpetian G., and Milimonfared J. Reliability Comparison of FuelCell DC-DC Converter in Two Cases of Using IPM Switch and Paralleling MOSFETs. Power electronics Specialists Conference, 2008, pp. 3723-3727, Rhodes, Greece, June.
9. Zhou L., Smedley, K. Reliability comparison of multilevel inverters for motor drive. Power & Energy Society General Meeting, 2009, pp. 1-7. 2009, Calgary, Canada, July.
10. Arifujjaman, M.; Iqbal, M., Quaicoe, J. A Comparative Study of the Reliability of the Power Electronics in Grid Connected Small Wind Turbine Systems. Canadian Conference on Electrical and Computer Engineering, 2009, pp. 394–397, St. John, Canada, May.
11. A. Ranjbar, B. Abdi, G. Gharehpetian, and B. Fahimi. Reliability Assessment of Single-Stage/TwoStage PFC Converters. International ConferenceWorkshop on Compatibility and Power Electronics, 2009, pp. 253-257. Badajoz, Spain, May.
12. A. Abdi, J. Ranjbar, G. Milimonfared, G. Gharehpetian. Reliability Comparison of Boost PFC Converter in DCM and CCM Operating Modes, International Symposium on Power Electronics, Electrical Drives, Automation and Motion, 2008. pp. 939- 943, Naples, Italy, June.

13. Hirschmann D., Tissen D., Schröder S., and de Doncker R., Reliability Prediction for Inverters in Hybrid Electrical Vehicles. IEEE Transactions on Power Electronics, Vol. 22, No. 6, November 2007, pp. 2511- 2516.
14. Space product assurance. Derating - EEE components (ECSS-Q-ST-30-11C Rev.1 - PR-Draft 1). European Space Agency, Requirements & Standards Division. The Netherlands, 2010.
15. Military Handbook MIL HDBK 217F: Reliability Prediction of Electronic Equipment. USA Department of Defense. Washington, USA, 1991. December. Available at http://snebulos.mit.edu/projects/reference/MILSTD/MIL-HDBK-217F-Notice2.pdf.
16. C. Aguilar Castillo, García Beltrán, C. D. and Morcillo Herrera, C., "Digitally Controlled Integrated Electronic Ballast with Dimming and Power Factor Correction Features " Journal of Applied Research and Technology vol. 8 No. 3, pp. 295-309, Diciembre 2010.
17. J. Rodríguez Reséndiz, Gutiérrez Villalobos, J. M., Duarte Correa, D., Mendiola Santibañez, J. D. and Santillán Méndez, I. M., "Design and Implementation of an Adjustable Speed Drive for Motion Control Applications " Journal of Applied Research and Technology, vol. Vol. 10 No.2, pp. 180-194, April 2012.
18. Pressman A., Billings K., Moray T. Switching Power Supply Design. Third edition. McGraw-Hill, 2009. pp. 45- 75.

CITATION

S.E. De León-Aldaco, J.H. Calleja-Gjumlich, H.R. Jiménez-Grajales, F.I. Chan-Puc, Experimental Assessment of Derating Guidelines Applied to Power Electronics Converters, Journal of Applied Research and Technology, Volume 11, Issue 1, February 2013, Pages 103-114, ISSN 1665-6423, http://dx.doi.org/10.1016/S1665-6423(13)71519-0.

CHAPTER 2

Paralleled DC-DC Power Converters Sliding Mode Control with Dual Stages Design

Bashar Khasawneh, Maha Sabra and Mohamed A. Zohdy

Department of Electrical and Computer Engineering, Oakland University, Rochester, Michigan, USA.

ABSTRACT

This paper proposes the new cascaded series parallel design for improved dynamic performance of DC-DC buck boost converters by a new Sliding Mode Control (SMC) method. The converter is controlled using Sliding Mode Control method that utilizes the converter's duty ratio to determine the skidding surface. System modeling and simulation results are presented. The results also showed an improved overall performance over typical PID controller, and there was no overshoot or settling time, tracking the desired output nicely. Improved converter performance and robustness were expected.

INTRODUCTION

Power electronics is the process and control of the flow of electric energy from a given source to a load in a shape that is optimally suited for its use. Modern electric systems demand high quality, reliable, efficient and light weight power supplies.

Higher power converters, such as the ones used in Electrified Vehicles (EVs) and aircraft power units, are also in increase demand as a result of the green acts taken on by many countries. The improvement in power switching devices such as the IGBTs and MOSFETs made the power electronics more appealing to many applications. Higher switching frequency, higher power capability and improved efficiency are the main reasons for the expanded application [1].

Paralleled DC-DC converters are used in telecommunication industry widely and operated under closed loop control to regulate the output voltage [2]. The nonlinearities of the DC-DC converters are due to interaction among the converter components and the switching nonlinearity behaviors. However, linearized average model is commonly used for converter analysis [3-9]. A drawback of the linearized model is that it cannot predict the dynamics of the converter in a saturation region [2]. For standalone converters, many nonlinear control approaches have been studied and proposed: Lyapunov based controllers [10-13], variable structure controllers VSC [14- 16], feedback linearization method [17] and Fuzzy logic controllers [18]. In [19], the sliding mode control is used and the results are shown for all types of converters (Buck, Boost and Buck-Boost). The results of this type of control where the state space averaging is applied to sliding mode control PWM converters, showed a good result but high ripple content on the buck-boost converter.

For paralleled DC-DC converters control, [2] proposed an Integral Variable Structure Control (IVSC) for N paralleled DC-DC converters. They emphasized that Variable Structure Control was a natural choice since the control and the plant are both discontinuous and have uncertainties. A fuzzy logic controller is proposed in [18] for the master slave concept of the DC-DC control, and the results showed robust and improved performance over classical linear control. In [20], the control is based on a steady-state DC model and a small-signal model, which showed that current sharing can be achieved without a dedicated current sharing controller. [21-43] explored many of the concepts mentioned above and some of the chaotic behaviors of the converters. The papers are added for the readers' convenience to further explore the concept of the DC-DC converter control and behavior.

In this paper, we will introduce a concept to improve the output quality and simplify the paralleled converters control to a simple single like converter. We will use a Sliding Mode Controller (SMC) with the Variable Structure Surface Design concept developed in [44] to control the converter. A comparison with a PID controller is presented.

MODELING OF BUCK BOOST CONVERTER

In this section a model for N paralleled buck boost converters to supply a common load is developed. Figure 1 shows a commonly used buck-boost configuration for paralleled converters supporting a single load.

This paper will introduce a new approach to controlling the converter to reduce the output harmonics and simplify the control to an equivalent of a single converter. First a single converter model is given by the following equations:

$$\dot{i}_l = \left(1 - s_1\right)\frac{v_c}{L} + \frac{S_1}{L}E \tag{1}$$

$$\dot{v}_c = -\left(1 - s_1\right)\frac{i_l}{C} - \frac{v_c}{RC} \tag{2}$$

The above equations show the model that the first section of the converter associated with, the switch s_1 takes on the values[0 1]. The current is the inductor current associated with the first inductor L_1 and v_c is the capacitor voltage as a result of the contribution of all other converters. For multi converters the output current i_o is the sum of all individual inductor currents and is given by the following equation.

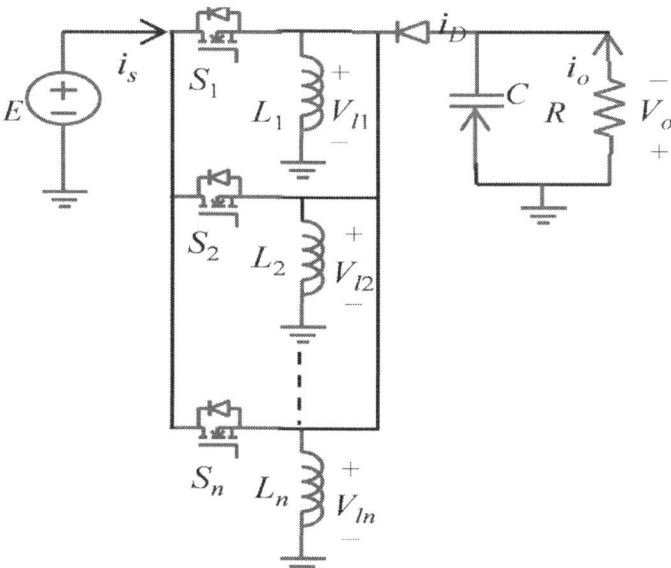

Figure 1. Paralleled buck-boost converter.

$$i_o = \sum_{n=1}^{m} i_{ln} - i_c \tag{3}$$

The states are the inductor currents and capacitor voltage as shown in Equation (4).

$$x = \left[i_{ln} v_c \right]^T, n = 1, 2, \cdots, m \tag{4}$$

m: number of converters Using Equation (3) and based on Equations (1) and (2) we can formulate the following relation. Given that the capacitor voltage is the same as the output voltage $V_c = V_o$.

The derivative of the output current given by Equation (3) is shown in Equation (5)

$$\dot{i}_o = \dot{i}_{l1} + \dot{i}_{l2} + \cdots + \dot{i}_{lm} - \dot{i}_c \tag{5}$$

Substituting Equation (1) into Equation (5) and collecting the terms we get Equation (6) interim of the derivative of the load and capacitor currents.

$$\dot{i}_o = \frac{v_c}{l} \left(\sum_{n=1}^{m} (1 - s_n) \right) + \frac{E}{l} \left(\sum_{n=1}^{m} (s_n) - \dot{i}_c \right) \tag{6}$$

where S_n is the state of the nth switch.

We determine from Figure 1 that the output voltage is equivalent to the capacitor voltage and the relation can be written in Equation (7)

$$v_c = v_o = i_o R \tag{7}$$

Taking the derivative of Equation (7) we get the following relation

$$\dot{v}_c = \frac{Rv_c}{l}\left(\sum_{n=1}^{m}(1-s_n)\right) + \frac{RE}{l}\left(\sum_{n=1}^{m}(s_n) - i_c\right)$$

$$= \frac{mRv_c}{l} + \sum_{n=1}^{m}s_n\left(\frac{ER}{l} - \frac{R}{l}v_c\right)$$

$$= \frac{mRv_c}{l} + \frac{R}{l}(E - v_c)\sum_{n=1}^{m}s_n$$

$$(8)$$

With the advancement in switches and their higher switching frequency capabilities we assume an infinite switching frequency. With this assumption we can assume that the rate of change in the capacitor voltage and current are zero, solving Equation (8) we get the following Equation (9)

$$\sum_{n=1}^{m}s_n = m\frac{v_c}{v_c - E}$$

$$(9)$$

It can be seen that the second term of Equation (9) is related to the duty ratio (d) for the buck boost converter. The duty ration is defined as the turn on time as a percentage of the period defined as T.

$$d = \frac{t_{on}}{T}$$

$$(10)$$

Substituting the expression for d into Equation (9) we determine the switching time as follows

$$t_{on} = T\left(\frac{s_1}{m} + \frac{s_2}{m} + \cdots + \frac{s_m}{m}\right)$$

$$(11)$$

The above equation shows the distribution of the switching time for the m converters, the sum of all the switches contribution is equivalent to the duty ratio or the required on time of an equivalent single switch to provide the desired output. The contribution of each leg is an equal portion of the required output voltage and current.

The control methodology can vary from PID, VSC, Fuzzy logic, SMC or other methods, but the end results is the same as to determine the turn ON time for the switches.

Next an improved performance of the converter to reduce harmonics content is presented. It is proposed to do a harmonic cancelation by shifting the switching sequence of the converters, by imposing t_{delay} on the subsequent converter legs using the following relation

$$t_{delay}(k) = t_d * \frac{(k-1)}{m}$$

(12)

t_d: Delayed time factor (e.g. 10 n sec)
k: The specific converter (1, 2, ……..m)
m: Total number of converters (2, 3…………)

DESIGN OF THE NEW CONVERTER CONTROL

In this section we propose a new converter control design to simplify the overall control and improve performance. In DC-DC converters control, the principle is to eliminate the error between the actual output and the desired output value. The control action is taking by switching the control device in this case a MOSFET to apply the input DC voltage to the converter periodically and proportional to the error. In this paper we propose separating the converter into two stages as discussed next.

The design consists of two cascaded stages where the first stage is the control stage shown in Figure 2 enclosed by the dashed lines and the second stage called the performance stage.

The performance stage represents m number of parallel converters to be controlled to reduce the harmonics contents of the output. This separation gives the designer the freedom to choose the switching frequency to optimize the performance of the converter. This frequency is independent of the control stage and can be optimized.

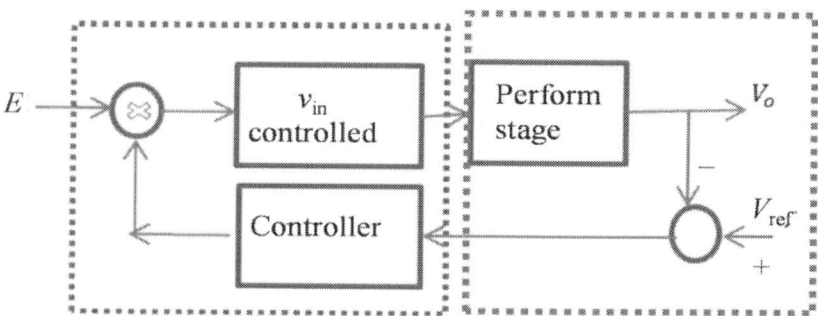

Figure 2. Two stages converter.

The converters switching time is determined in advance and the frequency can be optimized to reduce switching losses or improve EMI performance. The simulation will show the performance of the converter with synchronized switching time and also the enhanced performance with the shifted switching time as suggested above in Equation (12).

The control stage as shown enclosed in the dashed line in Figure 2 is used to control the input voltage to the paralleled converters. In previous converter designs the input voltage always assumed constant and the control was done by switching the converter at variable duty cycles to achieve the desired output values. In this paper we propose the idea of switching the output stage of the converter at an optimal switching frequency and duty cycle while controlling the input stage varying it to achieve the desired output value.

The control of the first stage through S_c, as shown in Figure 3, is designed with the assumption that the parallel converters have an equivalent inductor and capacitor values if the control method requires knowledge of the parameters. The control can be done using, Variable structure Control (VSC), SMC, where parameter sensitivity is not an issue, fuzzy logic, PID or any other method desired by implementation engineers as the problem is reduced to a single converter control.

SLIDING MODE CONTROL OF THE CONVERTER

In this section we will use sliding mode control to control the converter. The sliding surface used is a unique duty cycle dependent surface developed in [44]. The controller then derived using the sliding surface coefficients. In [44] a Variable Structure Control (VSC) surface design for DC-DC power converters is presented. The design was done for the Buck boost converter and can be extended and applied to all types using the methodology explained in [44].

Consider the linear time invariant system given by Equation (13).

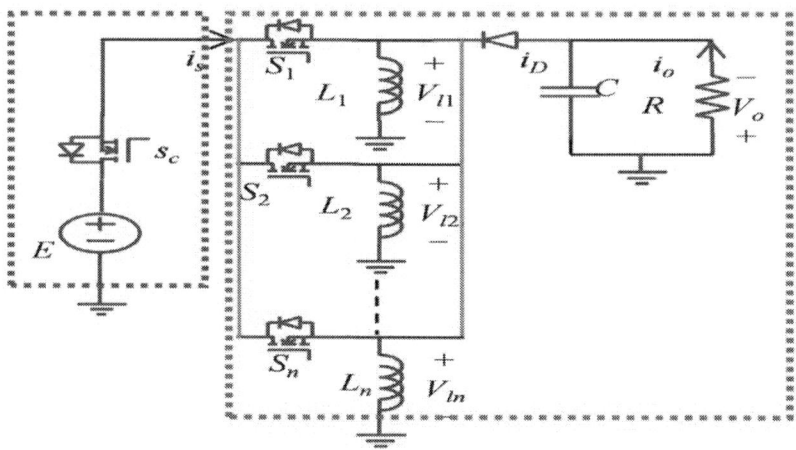

Figure 3. The proposed two stage converter.

$$\dot{x}(t) = Ax(t) + B(x)u(t) \tag{13}$$

where,

$$x = \begin{bmatrix} x_1 & x_2 \end{bmatrix} = \begin{bmatrix} i_l & vc \end{bmatrix}$$

$$A = \begin{bmatrix} 0 & 1/L \\ -1/C & -1/RC \end{bmatrix};$$

$$B = \begin{bmatrix} \dfrac{V_{in} - v_c}{L} \\ \dfrac{i_l}{C} \end{bmatrix}$$

The states are defined as the inductor current and the capacitor voltage which the same as the output voltage.

Given the surface as defined in [14,15].

$$s(x) = Cx \tag{14}$$

taking the derivative of (14).

$$\dot{s}(x) = C\dot{x}$$

(15)

Using Equation (13) to Equation (15) we get the equivalent control discussed by Utkin [16,29]. It is developed to derive the sliding mode equations into the manifold $s(x) = 0$ and then the solution to $s(x) = 0$ is called the equivalent control u_{eq}.

$$u_{eq} = -(CB)^{-1} CAx$$

(16)

The method in [44] defines the surface in term of the duty cycle and the parameters of the DC-DC converters as in Equation (17).

$$S(x) = Qx_{1e} + Wx_{2e}$$

(17)

Where

$$x_{1e} = \left(x_1 - x_{1ref}\right)$$
$$x_{2e} = \left(x_2 - x_{2ref}\right)$$

Are the errors between the actual values and the desired one. Taking the derivative of Equation (17) we find the value of C, the surface coefficients, as

$$\frac{dS}{dx} = C = [Q \quad W]$$

(18)

Using the value of C in Equation (18) and the given A and B coefficients of the system and evaluating Equation (16) to get the equivalent control u_{eq}.

$$u_{eq} = \frac{\dfrac{W}{C}x_1 - \left(\dfrac{Q}{L} - \dfrac{W}{RC}\right)x_2}{\dfrac{Q}{L}\left(v_{in} - x_2\right) + \dfrac{Q}{C}x_1}$$

(19)

The total control of the system consists of two components the equivalent control u_{eq} and corrective control u_c, where the equivalent control is used to reach the surface while the corrective is to keep the system on the surface.

$$u = u_{eq} + u_c$$

(20)

For system stability, we need to guarantee the system ends up and stays at the surface regardless of the initial conditions. Using the following Lyapunov function:

$$V(x) = \frac{1}{2}S(x)^2$$

(21)

We need to grantee that the derivative of Equation (21) is negative definite that is $\dot{V} < 0$ for all $S(x) \neq 0$, that from any initial condition. Taking the derivative of (21):

$$\dot{V} = S\dot{S} < 0$$

(22)

With

$$\frac{dS}{dx} = C$$

and

$$\dot{S} = CAx + CBu$$

to grantee Equation (22) holds the corrective control u_c is chosen as follow.

$$u_c = -K \operatorname{sgn}(S)$$

(23)

where K is appositive number. Hence the complete control is given by Equation (20) is:

$$u_c = -u = \frac{\dfrac{W}{C}x_1 - \left(\dfrac{Q}{L} - \dfrac{W}{RC}\right)x_2}{\dfrac{Q}{L}(v_{in} - x_2) + \dfrac{Q}{C}x_1} - K \operatorname{sgn}(S)$$

(24)

where Q and W are the coefficients derived in [2] and given below:

$$Q = \frac{L}{d}$$

$$W = -\left\{ L * \left(\frac{Jdl}{L} - \frac{Cd(d-1)}{LR} + \frac{Cdl(d-1)}{LR} \right) \right.$$

$$* \left(\frac{C(d-1)}{R} - J + \frac{(d-1)^2}{CJL} \right) \Bigg\} \Bigg/ \left\{ d \left(\frac{}{} \right) \right.$$

$$\left. * \left(\frac{Cd}{R} + \frac{dl(d-1)}{CJL} \right) * \left(\frac{(d-1)^2}{CL} - J + \frac{CJ(d-1)}{R} \right) \right\}$$

L : Inductor value

C : Capacitor value

R : Load Resistor Value

J : Associated Jordan value associated with the selected eigen values.

L and l: Arbitrary positive numbers

d : Duty ratio In the next section we will show the results of the application of the hyper plane coefficients in the proposed two stage converter and compare to the results to a conventional PID controller. Although a comparison will show an improvement over the conventional method in term of overshoot and settling time, the main purpose of the application to test the feasibility of the method developed in [44] and introduce the proposed two stage converter.

Figure 4 shows a single Buck Boost converter. Figure 5 shows paralleled buck boost converters. Both converters, the single or the paralleled are to regulate the output voltage to a desired value V_{ref} and provide loads with the desired currents. Figure 6 shows the sliding mode controller as given by (24).

In the next section will show and compare the results of using a single converter, parallel converters, paralleled converters with the enhanced mode and the developed hyper plain coefficients method [44].

RESULTS

The converter shown in Figure 4 represent a single Buck Boost converter based on the mathematical model described in (1) and (2). The converter was controlled using a PID control and sliding mode control with the developed hyper plain coefficients and simulated using the following circuit parameters:

$$l = 10 \text{ mh}, C = 10 \text{ μF}, R = 10 \text{ Ω}, E = v_{in} = 50 \text{ V}$$

In the second part of the simulation, the converters are paralleled and the new two mode design is implemented.

The same values as the single converter are used with a mismatch of 10% to represent a more realistic situation. However for sensitive applications more precise value components can be chosen.

Figure 5 shows the converters constructing a paralleled structure for the benefits mentioned earlier in the previous sections. The last part of the simulation was done using the delayed switching time as in (12) to further improve the performance.

First part of the simulation is of the single converters with the values given above. The results are shown below in Figure 7. The second part is the simulation Figure 8 is for the paralleled converter with a fixed duty cycle for all the converters in the performance stage. The third part Figure 9 is the results for the paralleled converters with the proposed switching delay as suggested by Equation (12).

Figure 7 shows the output voltage of a single converter regulated at 80 volts, the output is clearly showing the mount of ripple on the output voltage exceeds 5 V. The increase in the amount of ripples on the system voltage would cause increase losses reducing efficiency and can contribute to the mechanical wear and tear of a system.

Figure 8 shows the improved performance of the system by reducing the output voltage ripple to less than 1 V. This improvement is important as to increase system efficiency and improve its performance. In systems such as electric motors for example, this improvement can translate into increased reliability and reduction in maintenance cost.

Figure 9 shows further improvement in the sense of ripple reduction and more importantly significant reduction in harmonic content of the output. The use of the delayed switching time in the performance stage as suggested by Equation (12) has shown a great deal of improvement.

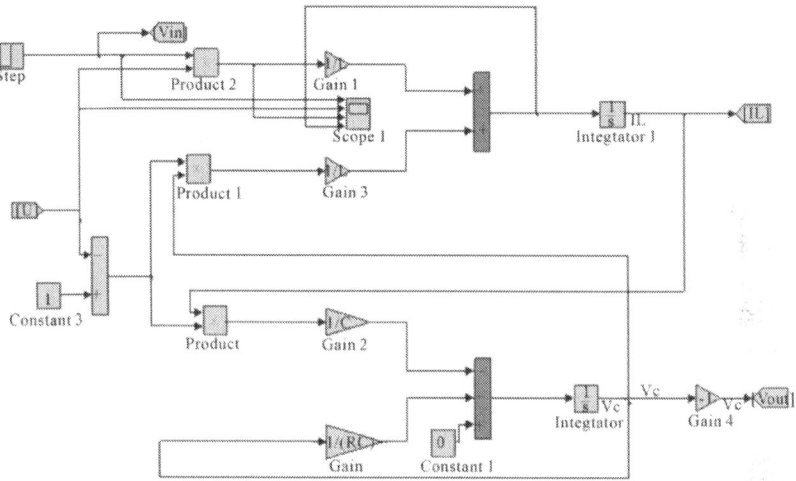

Figure 4. Simulink model of a single buck boost converter.

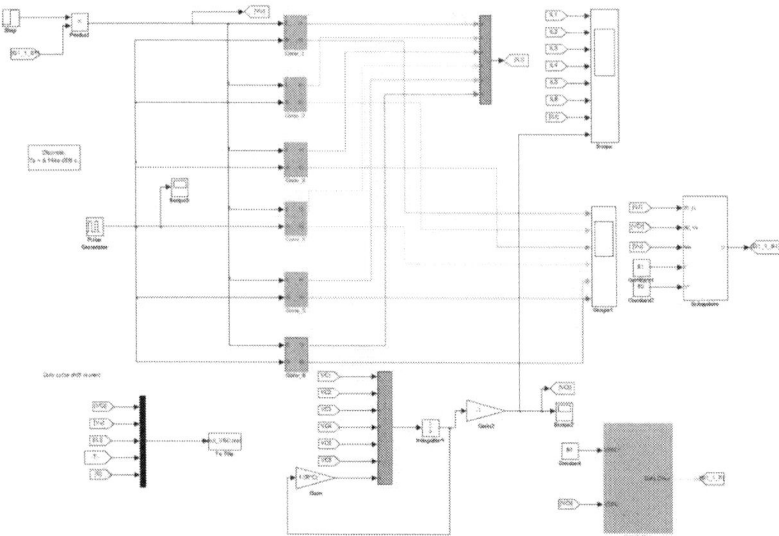

Figure 5. Simulink model of six paralleled buck boost converters.

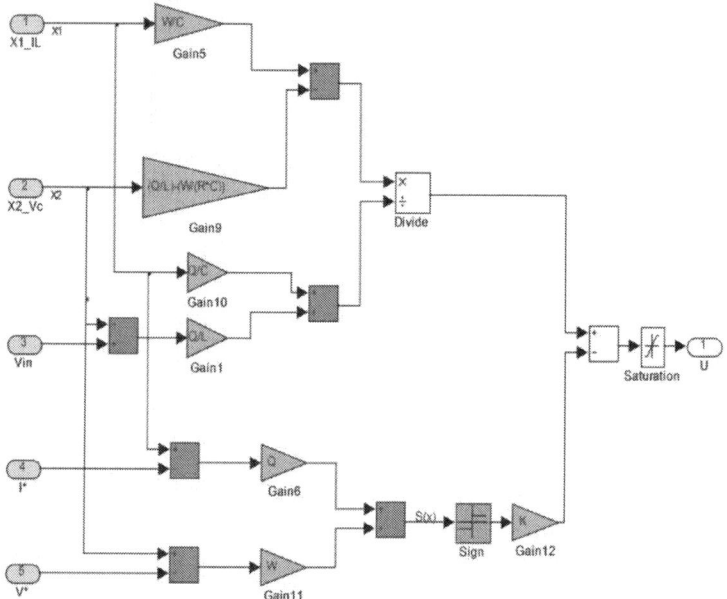

Figure 6. Simulink model of the control circuit as given in Equation (24).

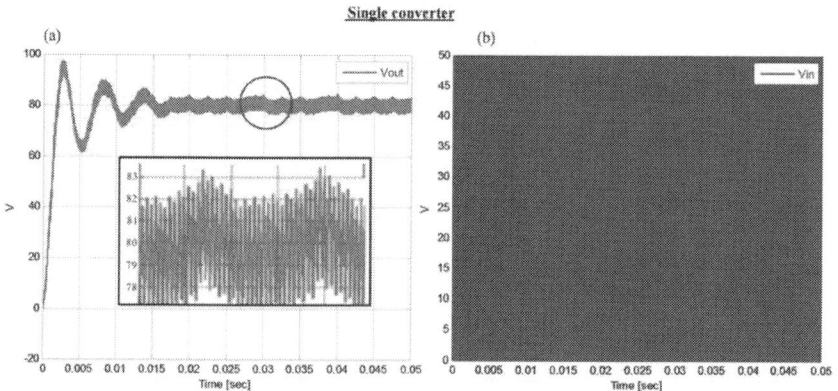

Figure 7. Output voltage, switched input voltage and a close-up of the output voltage of a single buck boost converter.

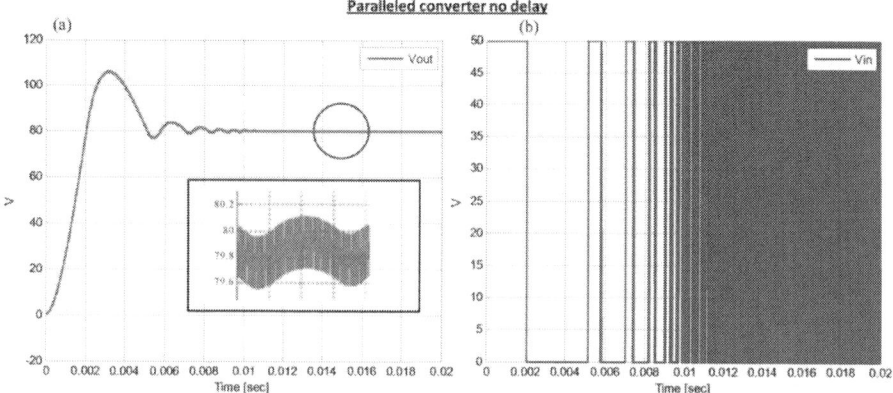

Figure 8. Output voltage, switched input respectively for paralleled converters with constant duty cycle at the performance mode.

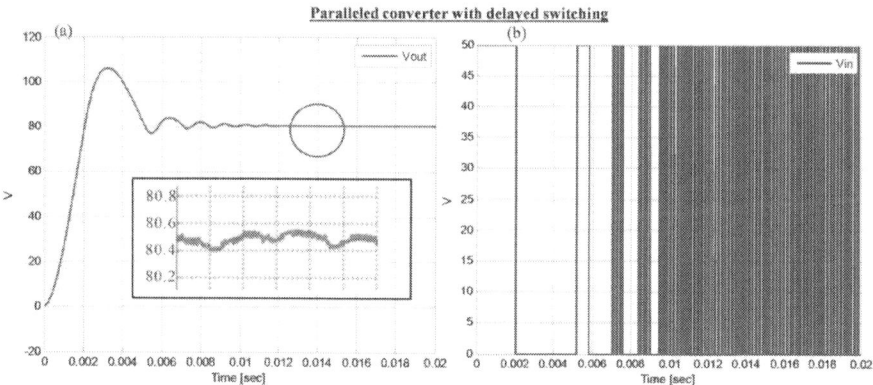

Figure 9. Output voltage, switched input voltage respectively for paralleled converters with enhanced duty cycle at the performance mode.

Examining the switched input voltages, it can be seen that the proposed two stage converter has reduced the switching frequency hence reducing the switching losses of the converter in the input stage. The performance stage switching frequency and duty cycle can be optimized to improve the overall efficiency of the system.

Figure 10 shows further improvement on the overall performance of the converters by reducing the ripple and improving the system response to eliminate any overshoot or oscillations. Figures 11(a)and (b) show he best performance with elimination of ripples and reduction in harmonics content, better system response as the desired voltage is being reached without the overshoot or oscillations.

CONCLUSIONS

This paper develops a paralleled Buck Boost DC-DC converter with two stages named the control and the performance stages. The control stage performed a manipulation of the input voltage whereas the performance stage was designed to improve further the quality of the output. The quality is measured by reducing the ripples, hence reducing power lost and increasing system reliability.

The converter is controlled using Sliding Mode Control method. The sliding surface in the controlled is developed to be dynamic and duty cycle dependent. This dynamic hyper plane or sliding surface showed an expected improved performance. The results also showed an improved overall performance over typical PID controller, and there was no overshoot or settling time, tracking the desired output nicely. The enhanced mode in combination with the sliding mode control has shown a reduction in harmonics in comparison to the SMC without the delayed enhanced mode.

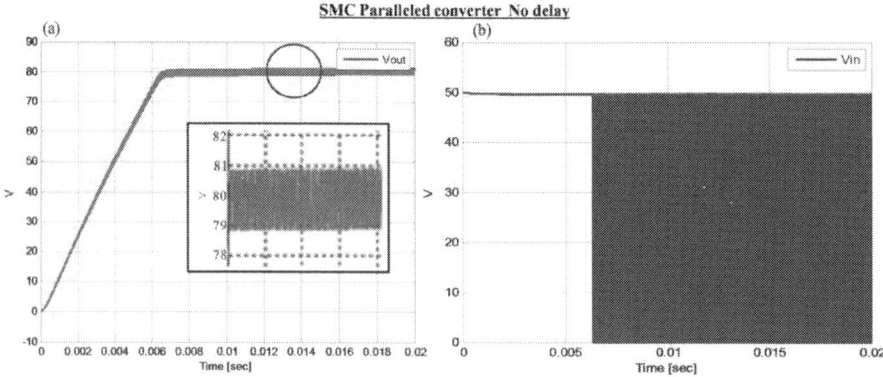

Figure 10. Output voltage, switched input voltage respectively for paralleled converters with constant duty cycle at the performance mode using SMC.

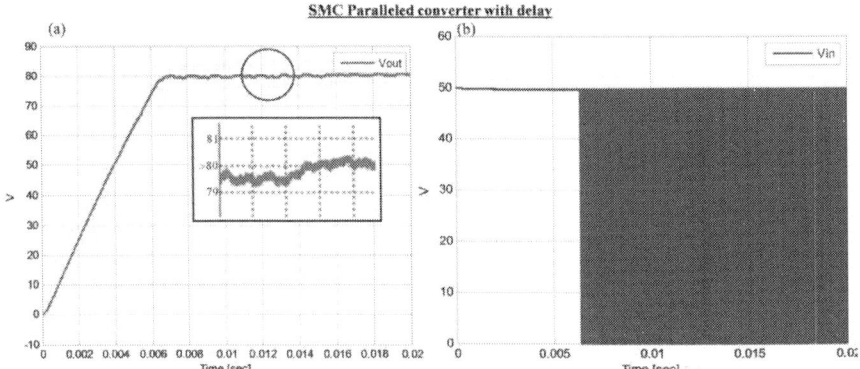

Figure 11. Output voltage, switched input voltage respectively for paralleled converters with enhanced duty cycle at the performance mode using SMC.

With the increasing demand of large power load and the development of distributed power supply system, the importance of research on paralleled power supply modules is increasing, while achieving an equivalent current sharing between the modules is the key element. In future work, the current-sharing control will be presented and compared with other existing ones.

REFERENCES

1. B. Khasawneh, M. Sabra and M. Zohdy, "Novel Operating Mode for DC-to-DC Converters in PHEVs," SAE Technical Paper 2013-01-1761, 2013.
2. S. K. Mazumder, A. H. Nayfeh and A. Borojevic, "Robust Control of Parallel DC-DC Buck Converters by Combining Integral-Variable-Structure and Multiple-Sliding-Surface Control Schemes," IEEE Transactions on Power Electronics, Vol. 17, No. 3, 2002, pp. 428-437.http://dx.doi.org/10.1109/TPEL.2002.1004251
3. T. Kohama, T. Ninomiya, M. Shoyama and F. Ihara, "Dynamic Analysis of Parallel-Module Converter System with Current Balance Controllers," 16th International Telecommunications Energy Conference, INTELEC'94, 1994, pp. 190-195.
4. V. J. Thottuvelil and G. C. Verghese, "Analysis and Control Design of Paralleled DC/DC Converters with Current Sharing," IEEE

Transactions on Power Electronics, Vol. 13, No. 4, 1998, pp. 635-644. http://dx.doi.org/10.1109/63.704129

5. K. Siri, C. Q. Lee and T. Wu, "Current Distribution Control for Parallel Connected Converters. I," IEEE Transactions on Aerospace and Electronic Systems, Vol. 28, No. 3, 1992, pp. 829-840. http://dx.doi.org/10.1109/7.256303

6. Qing Chen, "Stability Analysis of Paralleled Rectifier Systems," 17th International Telecommunications Energy Conference, INTELEC'95, 1995, pp. 35-40.

7. J. Rajagopalan, K. Xing, Y. Guo, F. C. Lee and B. Manners, "Modeling and Dynamic Analysis of Paralleled DC/ DC Converters with Master-Slave Current Sharing Control," 11th Annual Conference Proceedings on Applied Power Electronics Conference and Exposition, APEC'96, Vol. 2, 1996, pp. 678-684.

8. Y. Panov, J. Rajagopalan and F. C. Lee, "Analysis and Design of N Paralleled DC-DC Converters with MasterSlave Current-Sharing Control," 12th Annual Conference Proceedings on Applied Power Electronics Conference and Exposition, APEC'97, Vol. 1, 1997, pp. 436-442.

9. D. S. Garabandic and T. B. Petrovic, "Modeling Parallel Operating PWM DC/DC Power Supplies," IEEE Transactions on Industrial Electronics, Vol. 42, No. 5, 1995, pp. 545-551. http://dx.doi.org/10.1109/41.464619

10. P. V. Kokotovic, "The Joy of Feedback: Nonlinear and Adaptive," IEEE Control Systems, Vol. 12, No. 3, 1992, pp. 7-17. http://dx.doi.org/10.1109/37.165507

11. Krstic, Miroslav, I. Kanellakopoulos and P. V. Kokotovic, "Nonlinear and Adaptive Control Design," John Wiley & Sons, New York, 1995.

12. Khalil and K. Hassan, "Nonlinear Systems," Prentice Hall, Upper Saddle River, 2002.

13. Slotine, E. Jean-Jacques and W. P. Li, "Applied Nonlinear Control," Vol. 199, No. 1, Prentice Hall, Upper Saddle River, 1991.

14. M. Zohdy, M. S. Fadali and J. Liu, "Variable Structure Control Using System Decomposition," IEEE Transactions on Automatic Control, Vol. 37, No. 10, 1992, pp. 1514-1517. http://dx.doi.org/10.1109/9.256371

15. M. A. Zohdy, M. S. Fadali and J. Liu, "Variable Structure Dynamic Output Feedback," Proceedings of the 1995 American Control Conference, Vol. 2, 1995, pp. 1518-1522.http://dx.doi.org/10.1109/ACC.1995.521005

16. Utkin and I. Vadim, "Sliding Modes in Control and Optimization," Vol. 116, Springer-Verlag, Berlin, 1992.

17. Sastry and Shankar, "Nonlinear Systems: Analysis, Stability, and Control," Vol. 10, Springer, New York, 1999.

18. Jamshidi, Mohammad, et al., "Applications of Fuzzy Logic: Towards High Machine Intelligence Quotient Systems," Prentice-Hall Inc., Upper Saddle River, 1997.

19. I. Batarseh, K. Siri and H. Lee, "Investigation of the Output Droop Characteristics of Parallel-Connnected DC-DC Converters," 25th Annual IEEE Power Electronics Specialists Conference, PESC'94 Record., Vol. 2, 1994, pp. 1342-1351.

20. J. J. Shi, L. B. Zhou and X. N. He, "Common-Duty-Ratio Control of Input-Parallel Output-Parallel (IPOP) Connected DC-DC Converter Modules with Automatic Sharing of Currents," IEEE Transactions on Power Electronics, Vol. 27, No. 7, 2012, pp. 3277-3291.http://dx.doi.org/10.1109/TPEL.2011.2180541

21. Batarseh, K. Siri and H. Lee, "Investigation of the Output Droop Characteristics of Parallel-Connnected DC-DC Converters," 25th Annual IEEE Power Electronics Specialists Conference, PESC'94 Recordings, 1994, Vol. 2, pp. 1342- 1351.

22. A. G. Beccuti, M. Kvasnica, G. Papafotiou and M. Morari, "A Decentralized Explicit Predictive Control Paradigm for Parallelized DC-DC Circuits," IEEE Transactions on Control Systems Technology, Vol. 21, No. 1, 2013, pp. 136-148.http://dx.doi.org/10.1109/TCST.2011.2178071

23. A. A. Aboushady, K. H. Ahmed, S. J. Finney and B. W. Williams, "Linearized Large Signal Modeling, Analysis, and Control Design of Phase-Controlled Series-Parallel Resonant Converters Using State Feedback," IEEE Transactions on Power Electronics, Vol. 28, No. 8, 2013, pp. 3896-3911. http://dx.doi.org/10.1109/TPEL.2012.2231700

24. W. Qian, H. Y. Cha, F.-Z. Peng and L. M. Tolbert, "55-kW Variable 3X DC-DC Converter for Plug-In Hybrid Electric Vehicles," IEEE Transactions on Power Electronics, Vol. 27, No. 4, 2012, pp. 1668-1678. http://dx.doi.org/10.1109/TPEL.2011.2165559

25. Jamerson, Cliff and C. Mullett, "Paralleling Supplies via Various Droop Methods," High-Frequency Power Conversion Conference (HFPC) Proceedings, 1994, pp. 68- 76.

26. M. Grossoni and G. Cimador, "A Selective Supervision Device for Paralleling Operating ac/dc and dc/dc Converters," 5th International Telecommunications Energy Conference, INTELEC'83, 1983, pp. 587-593.

27. R.-H. Wu, T. Kohama, Y. Kodera, T. Ninomiya and F. Ihara, "Load-Current-Sharing Control for Parallel Operation of DC-to-DC Converters," 24th Annual IEEE Power Electronics Specialists Conference, PESC'93 Recordings, 1993, pp. 101-107.

28. Y. Panov, J. Rajagopalan and F. C. Lee, "Design-Oriented Analysis of Paralleled dc-dc Converters with Master-Slave Current-Sharing Control," Proceedings of the Annual Seminar of Virginia Power Electronics Seminar, 1996, pp. 83-89.

29. V. I. Utkin, J. Guldner and J. Shi, "Sliding Mode Control in Electromechanical Systems (Vol. 9)," CRC Press, Boca Raton, 1999.

30. Freeman, A. Randy and P. V. Kokotovic, "Robust Nonlinear Control Design: State-Space and Lyapunov Techniques," Springer, Berlin, 2008.

31. Z. H. Qu, "Robust Control of Nonlinear Uncertain Systems," John Wiley & Sons, Inc., Hoboken, 1998.

32. M. M. Gupta and T. Yamakawa, "Fuzzy Computing: Theory, Hardware, and Applications," North-Holland, 1988.

33. Kandel, Abraham, and G. Langholz, "Fuzzy Control Systems," CRC Press, Boca Raton, 1994.

34. E. Fossas and G. Olivar, "Study of Chaos in the Buck Converter," IEEE Transactions on Circuits and Systems I: Fundamental Theory and Applications, Vol. 43, No. 1, 1996, pp. 13-25.

35. M. Grossoni and G. Cimador, "A Selective Supervision Device for Paralleling Operating AC/DC and DC/DC Converters," 5th International Telecommunications Energy Conference, INTELEC'83. 1983, pp. 587-593.

36. S. Gutman, "Uncertain Dynamical Systems—A Lyapunov Min-Max Approach," IEEE Transactions on Automatic Control, Vol. 24, No. 3, 1979, pp. 437-443.http://dx.doi.org/10.1109/TAC.1979.1102073

37. D. C. Hamill, J. H. B. Deane and D. J. Jefferies, "Modeling of Chaotic DC-DC Converters by Iterated Nonlinear Mappings," IEEE Transactions on Power Electronics, Vol. 7, No. 1, 1992, pp. 25-36. http://dx.doi.org/10.1109/63.124574

38. S. K. Mazumder, A. H. Nayfeh and D. Boroyevich, "Theoretical and Experimental Investigation of the Fastand Slow-Scale Instabilities of a DC-DC Converter," IEEE Transactions on Power Electronics, Vol. 16, No. 2, 2001, pp. 201-216.http://dx.doi.org/10.1109/63.911144

39. C. K. Tse, "Flip Bifurcation and Chaos in Three-State Boost Switching Regulators," IEEE Transactions on Circuits and Systems I: Fundamental Theory and Applications, Vol. 41, No. 1, 1994, pp. 16-23.

40. J. R. Wood, "Chaos: A Real Phenomenon in Power Electronics," Proceedings of 4th Annual IEEE on Applied Power Electronics Conference and Exposition, Baltimore, 13-17 March 1989, pp. 115-124. http://dx.doi.org/10.1109/APEC.1989.36959

41. R.-H. Wu, T. Kohama, Y. Kodera, T. Ninomiya and F. Ihara, "Load-Current-Sharing Control for Parallel Operation of DC-to-DC Converters," 24th Annual IEEE Power Electronics Specialists Conference, Seattle, 20-24 June 1993, pp. 101-107.

42. B. Tomescu and H. F. VanLandingham, "Improved LargeSignal Performance of Paralleled DC-DC Converters Current Sharing," IEEE Transactions on Power Electronics, Vol. 14, No. 3, 1999, pp. 573-577. http://dx.doi.org/10.1109/63.761701

43. R. Marino and P. Tomei, "Nonlinear Control Design: Geometric, Adaptive and Robust," Prentice Hall International (UK) Ltd., 1996.

44. B. Khasawneh, M. Sabra and M. Zohdy, "Power Converters Variable Structure Control Surface Design," 3rd International Conference on Electric Power and Energy Conversion Systems (EPECS'13), August 2013, Article ID: 167-55397.

CITATION

Khasawneh, B. , Sabra, M. and Zohdy, M. (2014) Paralleled DC-DC Power Converters Sliding Mode Control with Dual Stages Design. Journal of Power and Energy Engineering, 2, 1-10. doi: 10.4236/jpee.2014.22001.

CHAPTER 3

Review on Electrical Control Strategies for Wave Energy Converting Systems

Yue Hong[1], Rafael Waters[1,2], Cecilia Boström[1], Mikael Eriksson[2], Jens Engström[1], Mats Leijon[1,2]

[1] Division for Electricity, Uppsala University, SE-751 21 Uppsala, Sweden
[2] Seabased AB, SE-751 21 Uppsala, Sweden

ABSTRACT

Renewable energy techniques are now gaining more and more attention as the years pass by, not only because of the threat of climate change but also, e.g. due to serious pollution problems in some countries and because the renewable energy technologies have matured and can be depended upon an increasing degree. The energy from ocean waves bares tremendous potential as a source of renewable energy, and the related technologies have continually been improved during the last decades. In this paper, different types of wave energy converters are classified by their mechanical structure and how they absorb energy from ocean waves. The paper presents a review of strategies for electrical control of wave energy converters as well as energy storage techniques. Strategies of electrical control are used to achieve a higher energy absorption, and they are also of interest because of the large variety among different strategies. Furthermore, the control strategies strongly affect the complexity of both the mechanical and the electrical system, thus not only impacting energy absorption but also robustness, survivability, maintenance requirements and thus in the end the cost of electricity from ocean waves.

INTRODUCTION

Oceans cover two thirds of the earth's surface and large amounts of energy is contained within its motion. This vast source of renewable energy has the potential of meeting an important part of the demand for non-polluting electricity for mankind.

Since the 1970s, following the oil crisis, research on renewable energy has gained increasing support and this has also affected the research on wave energy. So far, many types of wave energy converters (WECs) have been invented, developed and tested in small and large scale experiments.

When the electricity production from WECs depends on and varies synchronously with the wave movement then the amplitude and frequency of the converted electricity will naturally vary dramatically during each wave period. This electricity is incompatible with the electric grid since electricity supplied to the grid has to have a voltage that is of constant amplitude and frequency. Hence, some strategies of electrical control and storage as part of the energy conversion play a significant role in wave energy systems.

This paper is mainly concerned with electrical control strategies for wave energy conversion. In Section 2 a short introduction to wave energy absorption is presented. The purpose of this section is to help the uninitiated reader to understand what the control strategies are trying to achieve in terms of the interaction with, and absorption of energy from the wave. In order to facilitate relevant comparisons among control methods, the different types of wave energy converters are classified with respect to their mechanical structures in Section 3. In Section 4, electrical control strategies with devices as examples are discussed. Section 5 presents energy storage strategies that are used during the energy absorption.

This paper focuses on the control strategies and only provides a brief overview of existing wave energy technologies, for a more thorough review and classification of wave energy converters see [10]. It is necessary to mention that for various reasons, such as mechanical problems or lack of financial support, not all the control strategies have been experimentally verified as proposed and studied in their academic publications.

ENERGY ABSORPTION

Oceans transport huge amounts of energy, this energy is transported in form of polychromatic waves. An ocean wave is water particles moving in elliptical orbits, where the radius decreases with the water depth [1]. The power transport per unit width of wave front is given by the hydrodynamic pressure and the water particle velocity, according to

$$J = \int_{-z}^{0} p_i v_i \, dz \left[\frac{W}{m} \right] \tag{1}$$

The integral is calculated from bottom to still water level. An object or something else interacting with this incident wave will create a radiated and a diffracted wave that will change shape of the incident wave. To be able to take away energy from this incident wave this "object" has to interact in an advantage way to this wave. Being able to absorb energy from the waves, radiated and diffracted waves has to be created to interfere destructively with incident waves. A good wave absorber has to be a good wave maker[2]. Assume an arbitrary shaped area of the ocean surface (Fig. 1). The energy flux into this volume is denoted as E_i, and the energy flux out from the area is denoted E_o. According to energy balance the possible absorbable energy from the waves in the given volume is given by $E_i = E_o + E_{abs}$. Being able to absorb energy an outgoing wave has to be created by the absorber.

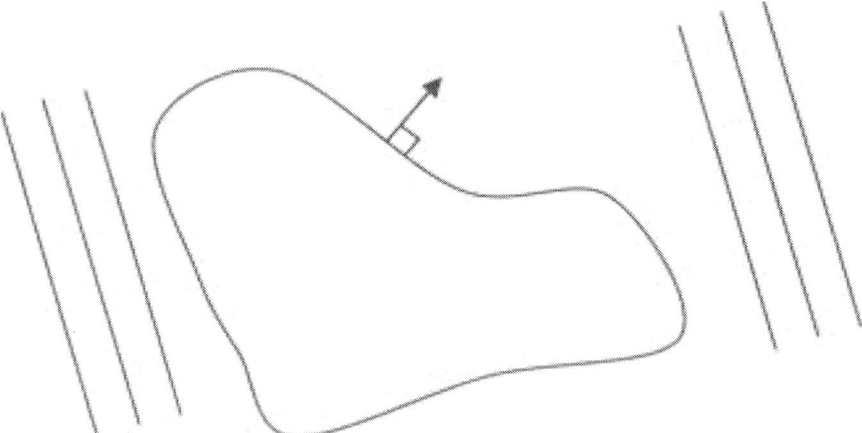

Figure 1. An arbitrary area of the ocean surface.

Assume an undisturbed incident wave has the water particle velocity, $v_i = (v_{i,x}, v_{i,y}, v_{i,z})$ and a hydrodynamical pressure p_o. The same for the diffracted/radiated wave from the wave absorber unit inside the volume, i.e. the outgoing wave from a wave absorber unit have water particle velocity, $v_o = (v_x, v_y, v_z)$, and a hydrodynamical pressure p_o. The total net flux of the energy in an arbitrary volume of the ocean during a time $2t$ is then given by

$$E_{abs} = \int_{-t}^{t} \oint_S \int_{-z}^{0} (p_i - p_o)(v_i - v_o) \cdot \hat{n} \, dz \, ds \, dt \qquad (2)$$

Note that for an undisturbed wave the absorbed energy is zero, because the incoming energy to the volume is the same as the outgoing energy.

In order to maximize the energy absorption from the ocean wave, Eq. (2) has to be maximized. It can be seen in Eq. (2) that the interaction is not momentarily, to maximize outtake of energy the incident wave has to be known in advance.

Absorption is about to create a wave with right amplitude, frequency and phase angle to cancel out incident wave.

CLASSIFICATION OF WAVE ENERGY CONVERTERS BASED ON MECHANICAL DESIGN

Several systems for the classification of WECs have been proposed through the years, ranging from detailed to rough, see [3] and [4]. For simplicity we have chosen a basic separation into three types: Oscillating Water Column devices (OWCs), overtopping devices, and attenuators (Fig. 2). OWCs and overtopping devices are both available for offshore and inshore installations. The attenuators are predominantly offshore devices.

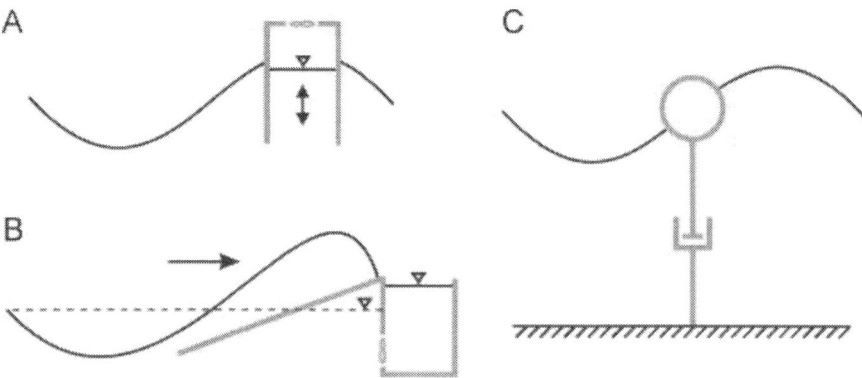

Figure 2. Mechanical design on three types of WECs [5]: (A) Oscillating water column device; (B) Overtopping device; (C) Attenuator.

Oscillating Water Column

The OWC (Figs. 2(A) and 3) device consists of a partially submerged chamber with a water column that rises and falls in response to the pressure from ocean waves [6]. Both the upwards and downwards movement of the column drive air through a turbine and the turbine drives a generator for electricity production [7].

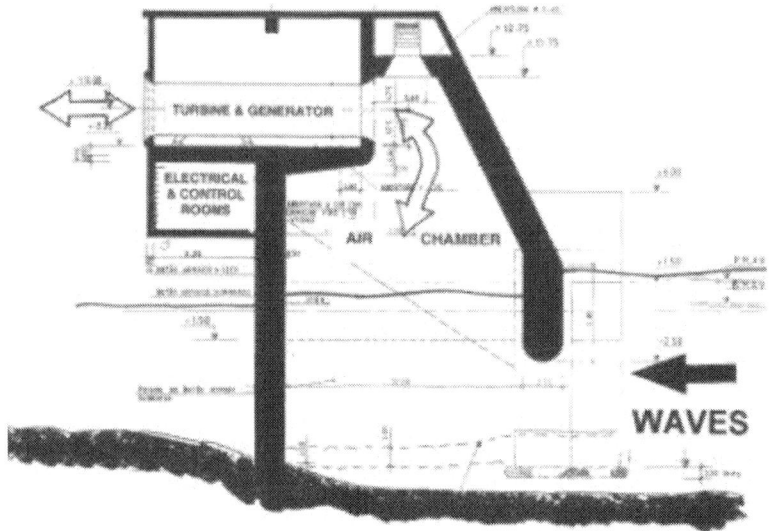

Figure 3. The Pico Plant on Azores Island in Portugal [15].

The water level in the column rises when a wave crest is pushing against the WEC. This increases the pressure in the device and forces air go through the turbine. Conversely, the pressure is lowered during a wave trough, resulting in air being pulled back through the generator. Both of these two processes drive the air turbines [8], and the rotational speed of the turbines depends on the air pressure in the column and the control system to the turbines.

Whistling Buoy, a device used as a navigational buoy, is patented by J. M. Courtney of New York, and is the earliest OWC device recorded in 19th centuries [9].

Yoshio Masuda in Japan invented a navigation buoy situated in Osaka Bay in 1947. It initially utilized an air turbine to supply electricity [10]. An advanced version, called the Kaimei, was built in 1976 and was owned by the Ryokuseisha Company. The Kaimei had an output power of about 70–500 W [11]. During 1978–1980, Kaimei Floating Platforms with rated power up to 125 kW was built and tested [12].

During the 1980s and 1990s, technologies for wave energy conversion were developed with increasing support as a result of the oil crisis that broke out in 1973. Large amounts of OWC devices from Norway, India, Japan and England, etc. were proposed at that time, expressing their own unique designs and advantages.

Queen's University of Belfast developed a device with an installed capacity of 75 kW positioned on Islay in 1991 [13]. This was followed by the OWC LIMPET, which was installed in 2000 and rated at 500 kW. In 2010, the European Wave Energy Pilot Plant(Fig. 3), which is located on the Azores [14], was operated with a rated power of 400 kW. Its installation demonstrated the technical viability of wave energy in a small island grid.

Overtopping Devices

Overtopping devices are partially submerged wave energy converter with reservoirs for capturing wave crests and water turbines to produce electricity [16]. The kinetic energy of the waves is converted to potential energy when incoming waves are led up a ramp and is collected in the reservoir (Fig. 2(B)). The water returns to the ocean from the reservoir through water turbines, thus utilizing the potential difference between the ocean and the reservoir to generate electricity.

A tapered channel wave power device, or Tapchan [17], utilizing this overtopping principle of operation, was installed on shore in the 1980s. The device was rated at 350 kW and located at Toftestallen, Norway [18]. Later on in 1998, a 1.1 MW Tapchan was started to be constructed on the Indonesian island of Java.

Wave Dragon installed a prototype in scale 1:4.5 (58×33 m, with a 28 m reflector and reservoir of 55 m^3) constructed in 2003 at Nissum Bredning in Denmark [19]. It had a rated power of 140 kW (7 turbines each with a rated power of is 20 kW). It became the first offshore overtopping device in the world (Fig. 4). Wave Dragon has started the development of 1.5 MW demonstrator offshore Hanstholm at the test center DanWEC, Denmark [20].

Figure 4. The Wave Dragon deployed and tested in the sea [20].

Attenuators

Attenuators are devices where the water physically pushes and induces motion in the WECs structure and energy is converted by dampening this motion [21]. Fig.2(C) represents the concept of attenuators. The devices are often constructions that float on and interact with the ocean waves without being physically fixed in place, e.g. the well-known Pelamis, but they can also be standing on the ocean floor or on land.

The Archimedes Wave Swing, AWS, is a submerged device [22] point absorber with a linear generator. A prototype was deployed off the northern coast of Portugal in 2000 and tested during 2004. It had a maximum power 2 MW [23].

Another example of a point absorbing attenuator is the WEC utilized in the Lysekil Project and developed at Uppsala University. Today nine 10–40 kW prototypes have been installed outside Lysekil off the west coast of Sweden [24]. Its linear generator is settled on the ocean floor and is driven via a connection line from the top of the generator to a buoy floating on the ocean surface (Fig. 5).

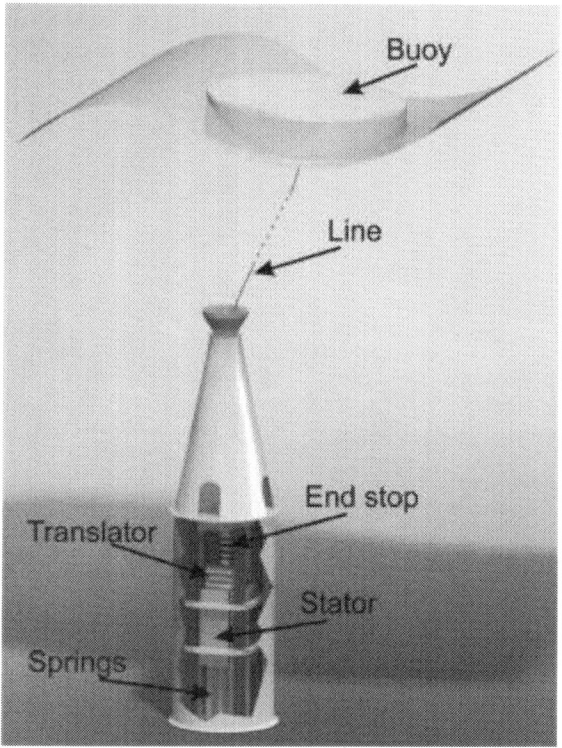

Figure 5. The WEC developed in the Lysekil Project [25].

In 2007, a prototype of a floating point absorber with a linear generator rated at 10 kW was developed by Oregon University and deployed off Newport in Oregon State in the US [26].

Wavebob is composed of two heaving buoys [27]: a torus of 14 m in diameter and a float linked to a submerged tank with a draught of 40 m.

The Wavebob belongs to the heaving buoy point absorber type. In 2007, Wavebob ltd. deployed its first test device in Ireland and planned to install a more advanced version with desired power production capacity in the order of several hundred kW off the coast of Portugal in 2012 [28].

Point Absorbers

Point absorbers are most commonly offshore devices that mainly utilize heave motion for energy absorption [29]. They are usually smaller size compared to other types of WECs. They can mathematically be regarded as a point-like object in the ocean due to their small size compared to the wave length of the wave from which they capture energy [30]. They are equally good at absorbing energy independent of the direction of the incoming waves. This type of wave energy converters are also categorized as oscillating body systems by Falcão.

Electrical controlling strategies devised for point absorbers vary depending on the mechanical design of the WEC. Hydraulic system and linear permanent magnet generator (LPMG) are, however, most common.

Hinged Attenuators

Hinged attenuators are made up of several body parts linked horizontally by universal joints which allow flexing in two directions. Taking Pelamis [31] (Fig. 6) as the most known example, it floats semi-submerged on the surface of the water and faces the direction of waves. As waves pass down the length of the machine and the sections bend in the water, the movement is converted into electricity via hydraulic systems housed inside each joint of the machine tubes, and power is transmitted to shore using standard subsea cables and equipment.

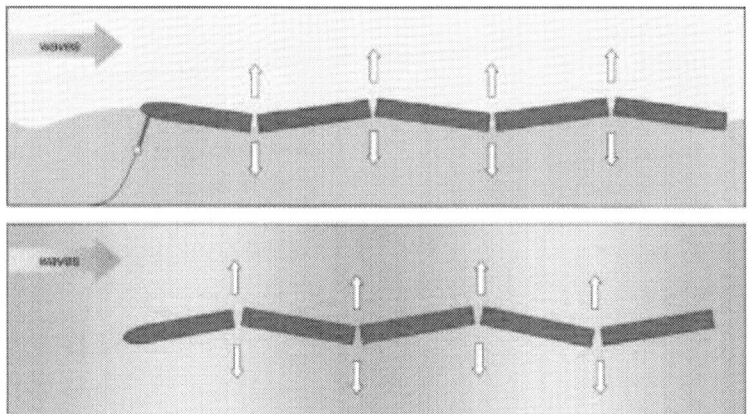

Figure 6. The concept of the Pelamis with hydraulic conversion system [32].

ELECTRICAL CONTROL STRATEGIES FOR DIFFERENT TYPES OF WECS

Control Strategies for OWCs

The performance of the OWCs lies in the combined efficiency of different stages (Fig. 7[33] proposes three types of turbine set) in the conversion process [34], they are (i) wave to pneumatic conversion in the capture chamber, (ii) pneumatic to mechanical conversion in the turbine, (iii) mechanical to electrical conversion in the generator.

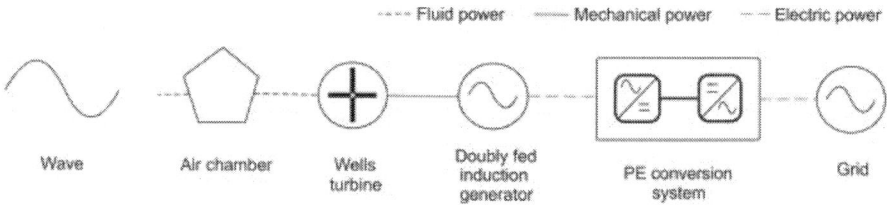

Figure 7. Technology used in a Pico OWC device to produce electricity from wave [91].

Due to challenges of the natural environment and device limitations, such as a mismatch between wave frequencies and the resonance frequency, variations in the air velocity, or variations in wave conditions, the efficiency of the overall system can be affected significantly [35]. For these reasons the combined efficiency has not been able to reach the theoretical values of efficiency of 70–80% in real operation. Additionally, the impact from flow oscillations onto the Wells turbine [36] also contributes to the low actual efficiency, with results that are far from the anticipated values.

In order to solve these problems, two control topologies are presented by Falcão [37]. These two topologies are utilized to solve the problems above to maximize the instantaneous power output of the wave energy converter: (i) the rotational speed control[38]; (ii) air-flow control strategy. They are designed to match the available wave level and the airflow control through the turbine, to prevent or reduce the aerodynamic blade losses at the turbine rotor blades.

A-Turbine Rotational Speed Control
The goal of the rotational speed control is to regulate the output power of the generator according to the error signal caused by abrupt changes in the fluid dynamics through the Wells turbines. Several strategies on the rotational speed control are proposed.

A variable frequency control strategy[40] is used as one strategy for the rotational speed control, which is a feedback control loop based on the cubic relationship between the frequency of the rotor in the generator and the instantaneous power of generator.

Falcão points out that improvement of the turbine efficiency [41] depends strongly on the control strategy to the instantaneous rotational speed. Three methods are proposed on controlling the turbine rotational speed within a limited range of flow conditions around the optimal efficiency point, these are (i) constant torque control depending on the power estimation [42], (ii) speed control to a reference rotational speed, with linear derivative function between torque and rotational speed, and (iii) control within acceptable oscillation [43]. Numerical simulations and analyses have been performed wherein the conclusion is reached that the third method (iii), electric power output controlled in the Programmable Logic Controller (PLC) [44], seems to be the optimal control method in the OWC device.

Srinivasa [45] also made comparison among three speed-control topologies to Wells turbine by simulations, which are (i) uncontrolled scheme, (ii) *V/f* (Velocity/Frequency)[46] controlled scheme on the stator and (iii) *V/f* controlled scheme on the rotor. Conclusion is that controlled scheme on the stator is the optimal topology, as power fluctuations will bring serious harm to the power factor in the grid under uncontrolled condition, and rotor circuit will absorb power under the rotor controlled condition.

Table 1 gives a summary of the control strategies to control the rotational speed of the turbines.

Table 1. Summarized control strategies of rotational speed control for OWCs.

Control strategy	Different topologies		Conclusion
Rotational speed control	(a)	Variable frequency control	It is a feedback control method for rotational speed control.
	(b)	(i) Constant torque control (ii) Speed control (iii) Speed control within acceptable oscillation	Numerical simulations show that (iii) is the optimal strategy for speed control.
	(c)	(i) Uncontrolled scheme (ii) V/f control on the stator (iii) V/f control on the rotor	Comparisons show controlled scheme on the stator is the optimal topology.

B-Airflow Control

As the efficiency of the Wells turbine depends strongly on the flow rate, flow oscillations will have a strong effect on the efficiency of OWC devices equipped with a Wells turbine. Thus, a method for controlling air valves in the chamber, in order to prevent excessive flow rate, has been proposed [47]. As Falcão describes in his article, two schemes of airflow control are compared and discussed: (i) valves mounted in parallel with the turbine; (ii) valves mounted in the turbine duct. According to the simulations, it is concluded that scheme (ii) is expected to achieve a better result [48] and [49].

In scheme (ii), error signals are induced by the difference between actual output power and reference output power. The reference power is set in accordance with the available pressure drop with the goal of maximizing the power output by avoiding the undesired stalling behavior. The control signal, attained by both the error signal and pressure regulation signal, drives the valve to adjust the pressure drop across the Wells turbine.

Furthermore, in practice, bypass valves with large capacity are required in order to limit the air flow through the turbine under extreme environmental conditions [50]. Table 2gives a summary of the control strategies of the airflow control for OWC devices.

Table 2. Summarized control strategies of airflow control for OWCs.

Control strategy	Different topologies	Conclusion
Airflow control	(i) valves mounted in parallel with the turbine (ii) valves mounted in series with the turbine duct	(ii) is better topology compared with (i) by avoiding the undesirable stalling to enhance the output power.

Typical OWC Type Wave Energy Converters
a. LIMPET

A turbine-generator system [51] is utilized and a torque control algorithm for the rotational speed control of the turbine is designed for the inverter drive connected to generators settled in LIMPET [52].

The system is composed of a simple switchboard feeding each generator system and connecting the plant to the grid via a 400/11 000 V transformer (Fig. 8). Each generator is controlled by the supervisory plant controller [53] that in turn controls an inverter drive connected to the machines.

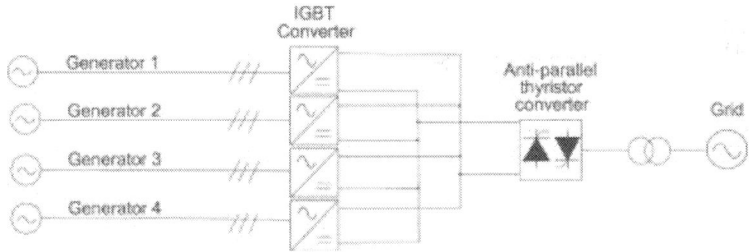

Figure 8. The electrical system of LIMPET [55].

Consisting of Insulated Gate Bipolar Transistor (IGBT) [54] based inverters and an anti-parallel Thyristor converter, the control system is used in LIMPET for adjusting the speed of the rotor is to optimize the relationship between its rotational speed and the velocity of the fluid flow. Three control strategies for rotational speed control are posited: (i) constant speed control; (ii) constant power control; (iii) constant torque control [55]. These three strategies are all implemented in the electric power interface in order to adjust the mean rotational speed to achieve the optimum speed by the generator's torque demand.

b. The Pico Plant

The Pico plant was equipped with a horizontal axis Wells turbine with fixed pitch blades with a rotational speed is in the range of 750–1500 rpm [56]. An asynchronous generator rated at 400 kW of the wound rotor induction type was adopted. The by-pass valve for air flow relief was designed and installed on the top of the air chamber so excessive energy can dissipate to the atmosphere when stormy weather might otherwise cause damage to the turbine-generator system.

The control and monitoring is achieved in a programmable logic controller (PLC), with the use of an interface that allows visualization of the main relevant parameters such as temperature, vibrations, the lubrication system of the turbine bearings, and the position of the valves, the power electronics and other electrical components.

Because of the complexity and the high cost of implementation, and because the environment is not suitable for installation, direct torque measurement at the turbine shaft was never installed [57]. As a result, the lack of instantaneous torque measurements has led to a lack of the possibility to apply control strategies to the generator.

Nevertheless, theoretical research has been pursued in spite of this [58] and [59]. Based on stochastic modeling [60], Falcão pointed out that the so-called cube-law utilized into a control algorithm of the relationship between the electromagnetic torque and the rotational speed, may be the optimal control strategy for the Pico Plant. Furthermore, different rotational speed control methods (with and without relief valve control) were compared and the results (Fig. 9), thick and thin lines represent the cube-law method without and with controlled valve respectively, while dashed lines represent a simplified method. The result [61] showed that proper control of the relief valves might lead to an increase of 37%.

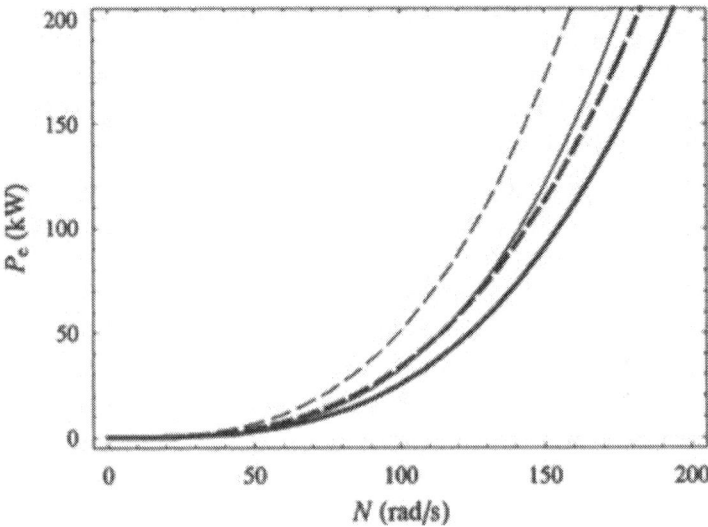

Figure 9. Comparison of rotational speed control methods (with and without relief valve control) [61].

Control Strategies for Overtopping Device

Tapchan is a typical demonstration of overtopping device in early years, however, the project stopped due to mechanical damage [62]. So far there is not any control information given out. In this section, the well-known Wave Dragon is used as the example to illustrate the electrical control to the WECs of overtopping devices. Fig. 10gives an illustration on the conversion stages from fluid power processed to electric power of one Wave Dragon.

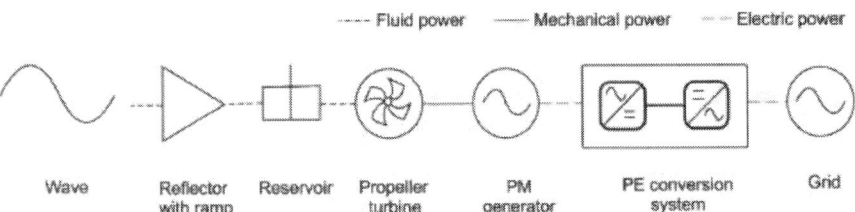

Figure 10. Technology used in Wave Dragon to produce electricity from waves [91].

The Wave dragon, is a slack moored [63] floating device consisting of two reflector arms and a central hull with a storage reservoir. The reflector arms focus the waves onto the doubly curved ramp on the front of the hull, where the waves run up and fill the reservoir. The water in the reservoir, which is situated above the mean sea level [64], is released back into the sea through a set of specially designed low head water turbines.

Direct power control, utilizing Direct Power and Torque Space Vector Modulation (DPTC-SVM) [66], is applied for the control of the converters. Direct Torque Control with Space Vector Modulation (DTC-SVM) is used on the Generator Side Converter to control the variable speed [67], while Direct Power Control with Space Vector Modulation (DPC-SVM) is used on Grid Side Converter with Pulse Width Modulation.

In order to decrease the actual DC-link voltage fluctuation during transients, the power is fed into the DC-link using a control device named Active Power Feed-Forward (APF).Fig. 11 represents the control system used in Wave Dragon. This approach shows high performance in stabilizing the speed of the generator, and it has shown to operate safely even during transient conditions.

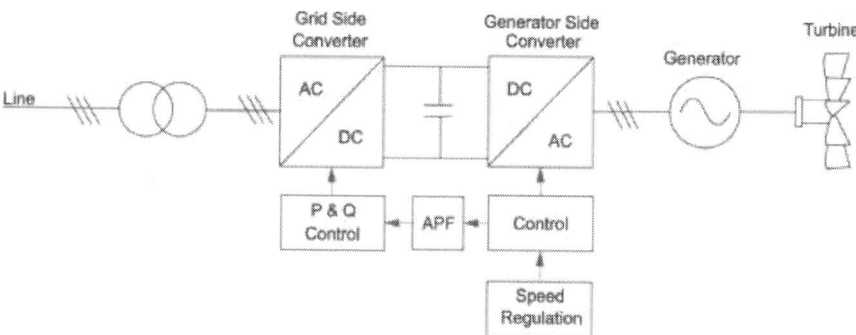

Figure 11. DPTC-SVM with APF for Active and Reactive Power Control (P&Q Control) [65].

A frequency power converter [68] is proposed and the current controller of AC/DC/AC converters is designed and simulated in dq0 frame, Fig. 12 shows the results of generated power and generator rotational speed, which are regulated to values with the maximum efficiency of turbine. Table 3 gives a summary of the control strategies designed on the converters of Wave Dragon.

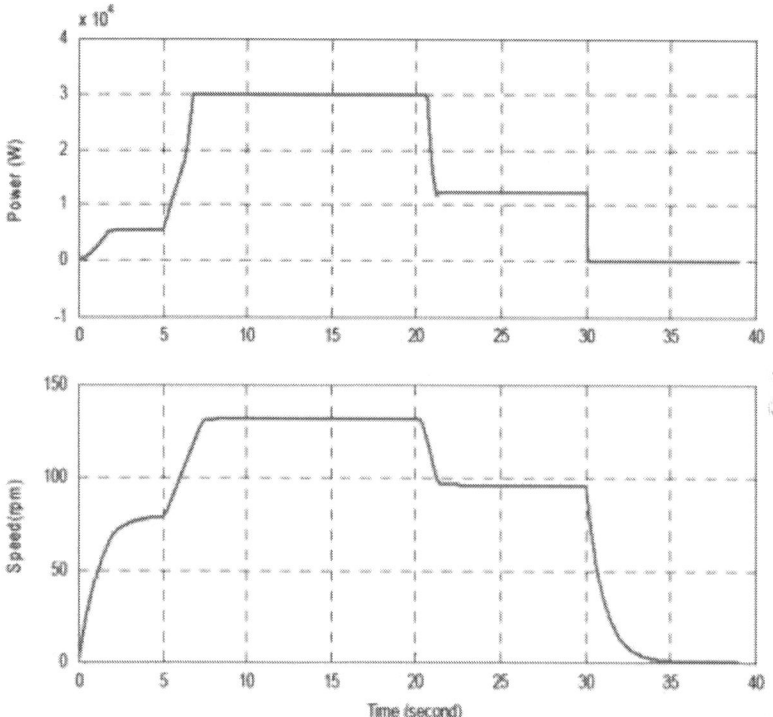

Figure 12. Output power from generator and generator speed with current control algorithm [67].

Table 3. Summarized control strategies of the examples for Overtopping system.

WECs	Control strategies on the converters	Conclusion
Wave Dragon	DTC-SVM on the generator side converter APF on the DC/DC control DPC-SVM on the grid side converter	The combined control is possible to achieve maximum efficiency of the turbine.

Control Strategies for Attenuators

A -Hydraulic System

Hydraulic systems are typically efficient energy conversion systems consisting of two energy conversion steps. The first step is that mechanical energy from waves is absorbed by a floating body and converted to hydraulic energy through a compression unit [69], which functions synchronously with the motion of floating body. Second, the generator or motor produces electrical energy in relation to the pressure difference resulting from the compression unit. A basic hydraulic system is explained in Fig. 13. Components of a basic compression unit include a hydraulic cylinder, two pressure accumulators and a valve-control system (Fig. 17). The hydraulic cylinder is used to transfer the motion of the floating body into that of direct-linked double-acting piston in the cylinder. Two chambers are connected respectively to a high pressure accumulator and to a low pressure accumulator, and the cylinder motion leads to a varied pressure difference that results in a torque on the generator/motor. Since unstable power output and energy loss is due to the existence of irregular waves, the valve-control technique is used to significantly smooth the power as well as for achieving energy storage for better efficiency.

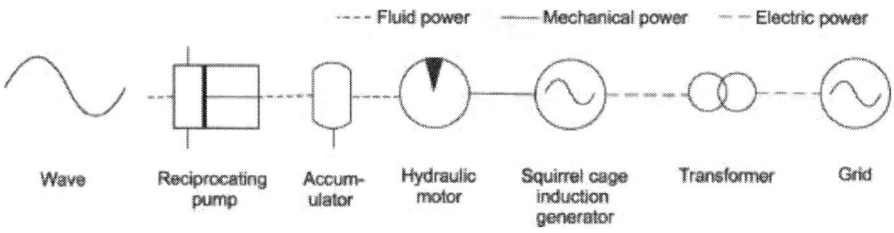

Figure 13. Technology used in a hydraulic system to produce electricity from wave [91].

In the following section, two examples of WECs for hydraulic system are illustrated with the control technologies, which are summarized in Table 4.

Table 4. Summarized control strategies of the examples for hydraulic system.

WECs	Control strategy	Description
Wavebob	Latching control	Control the damping force of the buoy system
Pelamis	Frequency tuning	Control the rotational speed and thus compressing motion in the cylinders

B-Typical WECs for Hydraulic System
a. Wavebob
In the hydraulic circuit of the Wavebob, two separate elevated pressure levels are set next to the biased low pressure (LP) level: the variable pressure (VP) and the high pressure (HP). They are established for de-

coupling of power output and input. The HP level is used to maintain the displacement control through HP motor, in order to maintain the working pressure for the hydraulic motor and to compensate the variations. The VP level is used to provide a defined damping force through the pumping module by defining the backpressure [70].

A Hydraulic Parallel Circuit (HPC) is preferred to maintain constant pressure and speed conditions within the limits set by the accumulator capacity. The AC generator is driven by two variable displacement motors—A HP motor and VP motor placed in parallel on a common shaft. While the VP motor displacement ramps up and down with the wave-dependent input flow, the HP motor is used to maintain a constant generator speed.

According to the structure of a HPC, latching control [71] is used as a method to control the damping force. The latching control strategy is achieved by two different control methods: VP motor control and HP motor control. Control of the VP motor is to control the damping force and pressure linearly proportional to the velocity of buoy system [72], while control of the HP motor is to maintain the accumulator pressure by adjusting the flow rate of the HP motor. Fig. 14 shows a schematic block of the function of the hydraulic system in Wavebob [73]. And in Fig. 15, the one on the top is the absorbed mechanical power by Wavebob, while the bottom one represents the output electrical power from a Power Take Off (PTO) system with non-linear damping control. The test results conclude that Wavebob succeeds to gain smooth output power under irregular wave conditions with the damping control.

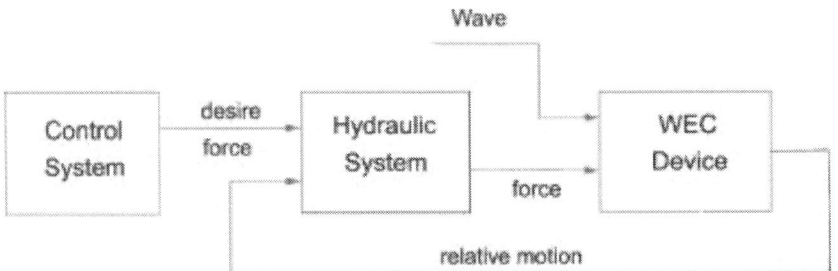

Figure 14. Schematical block diagram of the control to WEC of the Wavebob [73].

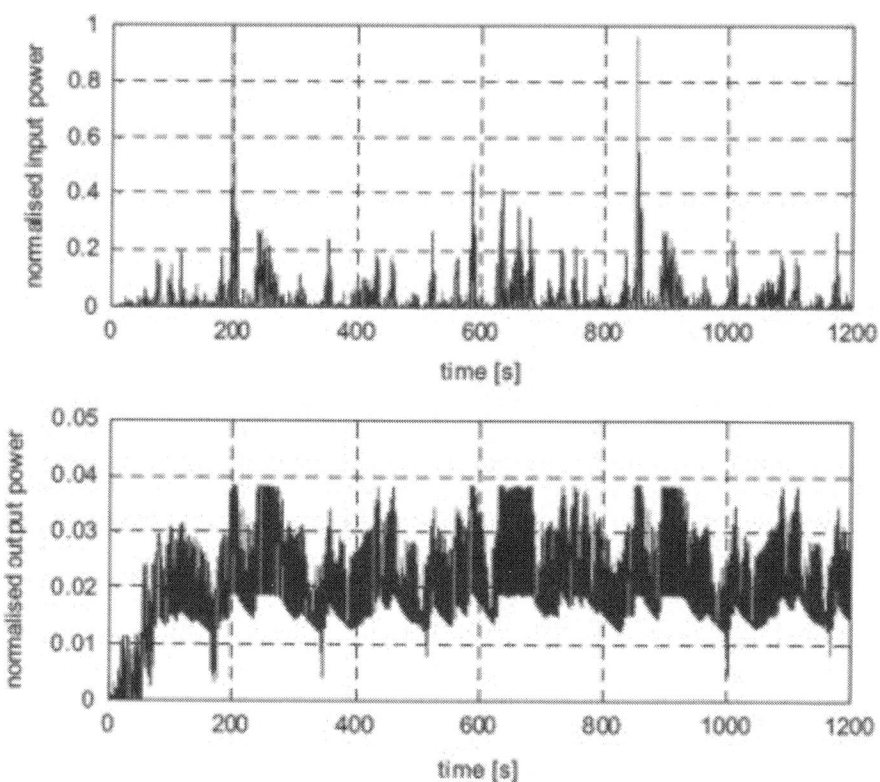

Figure 15. The absorbed power and the output electrical power from Wavebob [73].

In [69], a general hydraulic model is built theoretically attached with a point absorber WEC. It could also be fitted to the Wavebob, in spite of having been conceived for SEAREV. In the model, an extra accumulator is designed for energy storage [74]. Controlled valves with a PID controller are installed in each accumulator [75]. Fig. 16represents the output electrical power with different accumulators control method. When two accumulators are both working, the electrical power is enhanced by up to 50% in comparison with power with no accumulator control. In spite of the increase of the electrical power, this control strategy actually decreases the efficiency of the PTO.

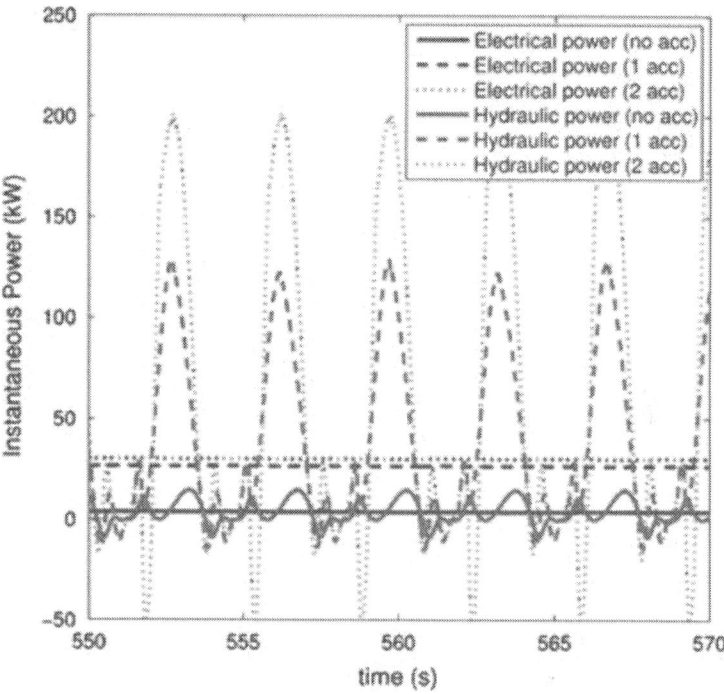

Figure 16. Output electrical power from the hydraulic system with extra accumulator control [69].

b. Pelamis

The Pelamis is a hinged wave energy converter consisting of cylinders connected by the joints aiming at pumping high pressure oil through hydraulic motors to drive electrical generators to realize electrical conversion [76]. The structure is stretched out along the direction of the moving waves [77], its snake-like motion is caused by the forces from waves and produces an angular relative motion between the cylinders which in turn leads to the pressurization and de-pressurization of the hydraulic system.

The hydraulic system is composed of two transmission parts. One is the primary transmission, in which energy from wave motion is converted into the hydraulic energy stored in the accumulators [78]. The secondary transmission is the conversion of the stored energy into electricity in the generators by use of hydraulic motors (Fig. 17).

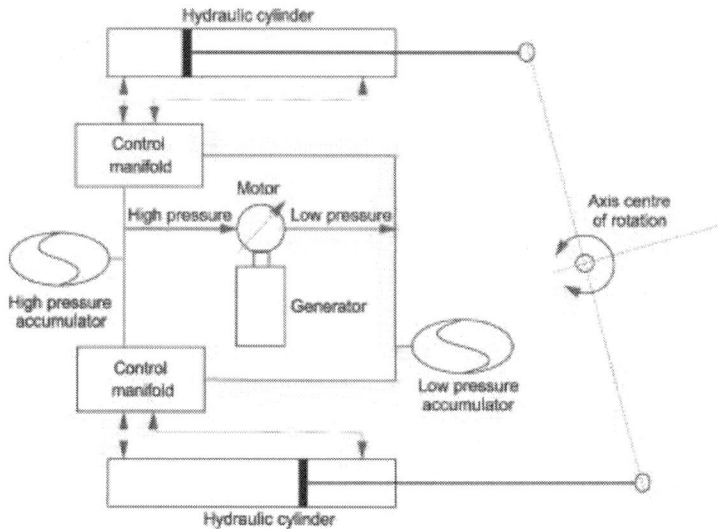

Figure 17. Simplified schematic system of the Pelamis [78].

The efficiency of Pelamis can be enhanced [79] by the frequency tuning, which is achieved by the control to the bias angle, leading to the control of rotational speed and thus compressing motion in the cylinders. The frequency tuning [80] mainly depends on the high-pressure fluid inside the hydraulic system. Therefore, electronically controlled valves are installed to control the flow of fluid and consequently affecting and controlling the angular relative cylinder motion of the Pelamis. Fig. 18 gives a simulated plot of absorbed and generated power from full-scale Pelamis operating under regular sea condition. The generated power shows that the smoothening effect from the secondary transmission is achieved comparing with the instantaneous power absorbed in the primary transmission. Additionally, Pelamis Wave Power Ltd. goes further with the theoretical research on maximum wave energy absorption by volume constraints, more details in [81].

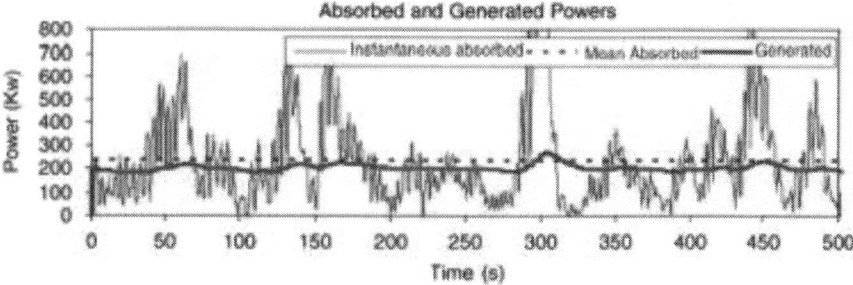

Figure 18. Absorbed and generated power from hydraulic system with active control strategy during continuous wave periods [78].

A-Direct Drive System

Direct drive system is a type of wave energy converter that generally has a more simple mechanical structure compared to that of hydraulic systems. It is employed in order to absorb the available wave energy more efficiently. The direct drive system with a linear generator [82] consists of a magnetic translator which is driven to reciprocate synchronously with the motion of a directly coupled buoy/floating body [83]. A simplified structure of direct drive conversion system is shown in Fig. 19. The result is directly induced three phase AC power without intermediate energy converting steps.

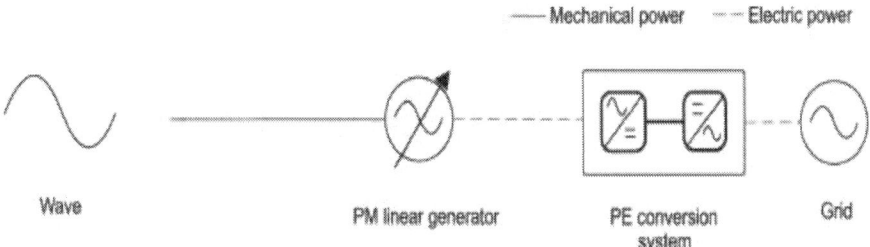

Figure 19. Technology used in a direct drive system to produce electricity from wave [91].

Directly driven systems have the advantage of not requiring an intermediate mechanical interface and thus avoiding the losses that take place in these devices (turbines and hydraulic motors) in other PTO systems. On the other hand, linear electrical generators for wave energy applications are in need of power electronics in order to convert the generated electricity to a form that is suitable for the electric grid.

Table 5 gives a summary of the control strategies used in the examples in the following section.

Table 5. Summarized control strategies of the examples for point absorbers.

WECs	Control strategy	Description
The Lysekil Project	Resonant circuit	To achieve electric resonance with the generator's winding
AWS	Feedback linearization control	To cancel the non-linear dynamics of the generator in order to achieve closed-loop linear control
	Reactive control	Control the output impedance to be equaled to the complex conjugate of the intrinsic impedance of generator
	Phase and amplitude control	Control the floater's vertical velocity in phase with wave excitation force, and the amplitude is also regulated
	Latching control	Control the water damper to prevent the floater from moving
	Stiffness and damping control	Current control in the dq0 frame to control the active and reactive power respective stiffness and damping factor
The Oregon University L10	Vector control of PADA system	Control the generated current's phase to be 90° ahead of the flux inside the generator

B-Typical WECs for Direct Drive System
a. The Lysekil Project

The linear generator developed in the Lysekil Project [84] is settled on the sea floor, via the connection line from the top of generator hinged to a buoy floating on the ocean surface [85], as illustrated in Fig. 5. So far, three different control strategies have been experimentally tested in the project. They are the passive diode rectification, DC control[86] and resonance circuit [87]. Future work will continue with the testing of marine substation [88]. During 2011 and 2012, a novel passive rectifier circuit was tested. The circuit can be seen in Fig. 20 and is a combination of a diode rectifier and resonance circuit with capacitors. The purpose with the circuit is to achieve a higher damping, e.g. power absorption, and at the same time maintain a high voltage output from the generator.

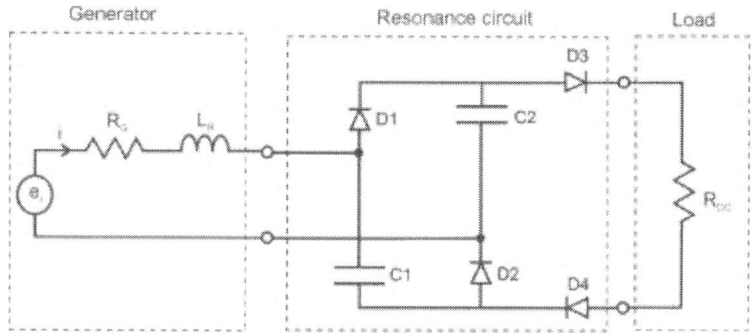

Figure 20. Installed electrical system in the measuring station [85].

The generator is connected to the rectified circuit [89] and in the experiments the translator's reciprocating motion was achieved with the help of a crane. The result from the experiment is presented in Fig. 21. The result [90] shows that the voltage across the capacitor C2, is increasing when the translator speed is decreasing to an operating frequency which is closed to the resonance frequency [91].

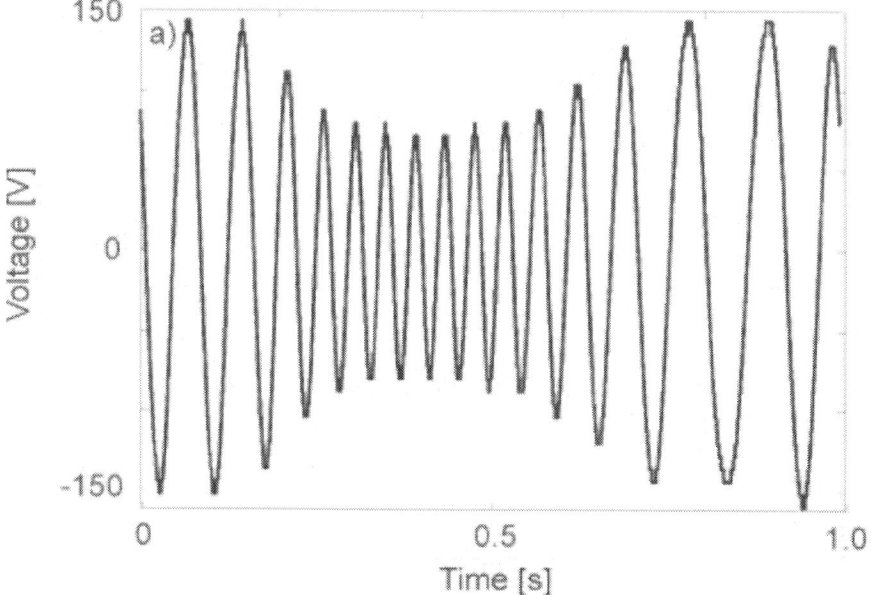

Figure 21. Measured voltage across capacitor, C2, result from [85].

b. Archimedes Wave Swing

The AWS consists of a submerged floater [92] that oscillates with the frequency of the ocean waves passing the device overhead. It is standing on a foundation fixed to the sea floor. Its reciprocating motion is driven by the varying pressure from the waves that act on an internal gas spring, causing a synchronous motion relative to the waves [93]. Since the wave energy is converted via the directly driven LPMG to produce electrical energy, strategies on converter control is of significance for extracting maximum energy from the waves.

Consisting of two voltage source converters (VSCs) and one capacitor, a full scale back-to-back converter is used for the converter system in AWS, with target of higher energy yielding and the converted power fluctuation smoothening [94].

The generator side converter for AC/DC conversion is the one of the VSCs connected to the stator of LPMG, while the grid side converter for DC/AC conversion is also the VSC linked to the grid on land [95]. As variable frequencies and magnitudes of power induced in the generator will cause dramatically fluctuations in the transmission, it is necessary to insert the storage unit—capacitor to smooth the ripples and guarantee the constant output of power to the grid.

Electrical control strategies to generator side converter of AWS with anticipation for energy absorption have been proposed. Feedback linearization control [96] was employed in the prototype of AWS. Its aim is to provide a control action that cancels the non-linear system dynamics [97] of the plant.

In spite of employing it into the realistic environment, obvious increase of the energy absorption is obtained by the feedback linearization strategies. Furthermore, several control strategies including phase control and amplitude control [98], reactive control, and latching control [99], [100] and [101] are proposed and relevant models are simulated comparatively. Fig. 22 shows the simulation result from database from different strategies and comparison with energy absorption without control is done [102]. From the result the feedback linearization is found to be the best control strategy, then following the latching control, the reactive control, the phase and amplitude control.

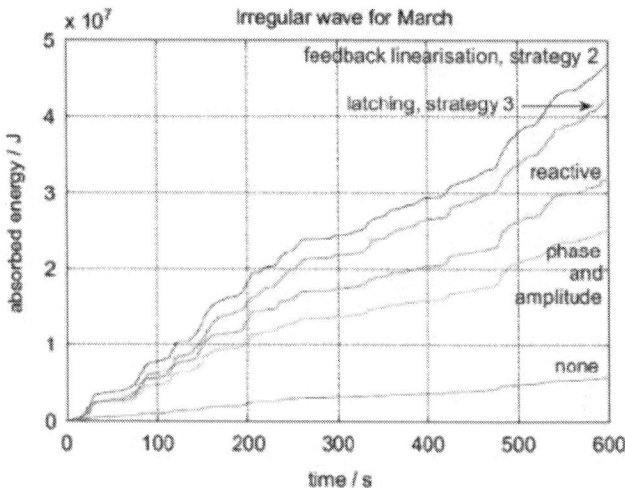

Figure 22. Results of the energy absorption in March with comparison among different control strategies [102].

Furthermore, a method of a dq0 model was proposed for dynamics and stability analysis of the stiffness and for the damping control on the generator instead of abc model that was used to control the prototype. Results (Fig. 23) from simulation show the proposed controllers are able to yield higher output voltage [103], and higher reactive power with an increase of 14.3%, and a higher sensation is as well achieved in tracking control reference value, even under situation of small disturbance.

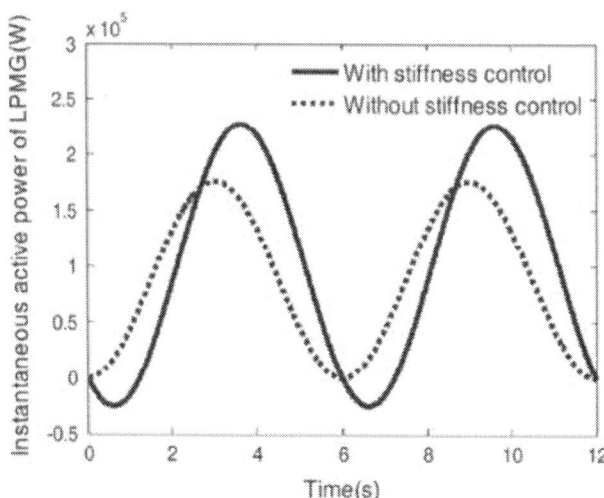

Figure 23. Instantaneous active power from LPMG with stiffness control and without stiffness control [39].

c. L10 of Oregon State University

The L10 consists of a deep-draught spar, restricting the device body to heaving motion. The motion of the linear generator is controlled by a stiff voltage applied to the terminals produced in an active control module.

The Power Analysis and Data Acquisition (PADA) system [104] for the electrical control to the generator is designed to fit in the WEC with rated power 30 kW [105], including functions of vector control [106], active and passive rectification [107] and [108], maximum power point tracking, and power control[109].

PADA system has two main power electronic components [110]: three-phase active rectified circuit and an output buck converter. The three-phase rectifier is connected to the generator and provides a path for the power flow into a common DC link. The output buck converter connected to the backside of the DC link provides control of the power flow to a fixed resistive load. The schematic system is described in Fig. 24, PWM controllers and phase controllers are comprised to control the IGBT Switch board on both side of the converter, aiming at regulating the energy absorption to maximum and to smoothen the power flow to the grid. All data from the voltage and current sensor to the DC link are sent through the remote monitoring.

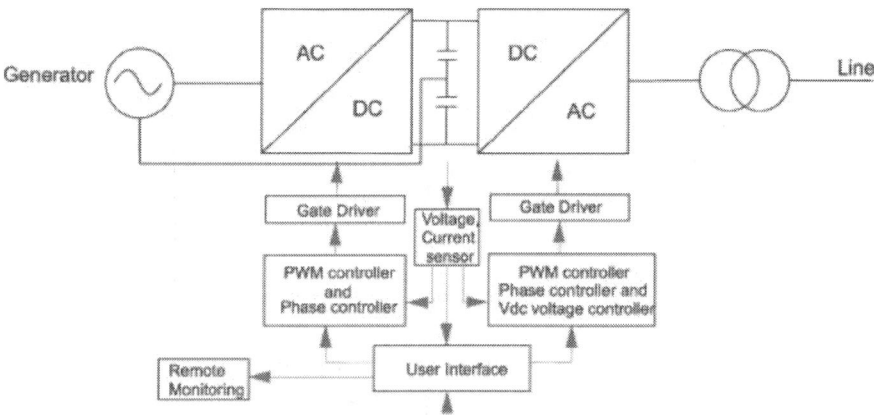

Figure 24. Schematic conversion system of the so-called L10 of Oregon University.

Test results with active rectifier with PADA system show that the DC link voltage is achieved to maintain the desired DC link voltage maximum of 600 V [111]. Fig. 25presents the result of generated power with active control during one hour session, showing that PADA system is promising to improve the power factor of the generator over that of a passive rectifier.

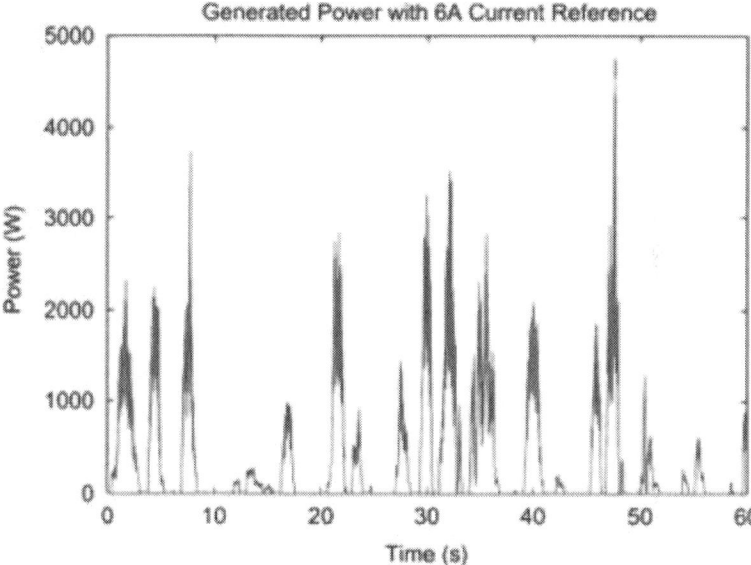

Figure 25. Generated power with current reference during 1 h session [104].

ENERGY STORAGE TECHNIQUES FOR WAVE ENERGY CONVERTERS

The output of energy from WECs varies dramatically due to the intermittent nature of the wave motion, both on the second scale and on the scale of hours and days. The resulting fluctuations of the harnessed energy affect the quality of the electricity supplied the grid, and increase the costs of the transmission. Because of this physical fact of the energy source, energy storage or buffer systems are often deemed as particular necessary for smoothening the variations and delivering high quality electrical power, and in order to provide stable electricity to grid. Short time scale variations of the input power from the WECs to the grid can be handled by utilizing energy storage systems as the examples described below.

Oscillating Water Column Devices

In the case of the demonstration wave energy plant in India [112], a high-speed flywheel type storage device was introduced along with an inverter to the three phase load for short duration energy storage. With the help of PID controller, the output load was switched to maintain the frequency and the voltage of AC supply within small variation, making it possible to gain balance between the input power from generator and output power to the load by adjusting the load value. Another solution is to place Superconducting Magnetic Energy Storage (SMES) across the DC link after the three-phase rectification with power electronic controller.

Hydraulic Systems

In the Pelamis' conversion system, short-term energy storage is provided by high pressure accumulators that help to deliver a smooth power flow to drive grid-connected electric generators [113]. Storage takes effect whenever there is a fluid exchange between the pressurizing chamber and the high pressure accumulators. The main problems contributed to the energy loss are related to mechanics, including the compressibility, bearing and seal friction of the hydraulic cylinders, and flow losses through valves and pipes.

Direct Drive Systems

Battery Energy Storage (BES) is proposed to be integrated into the electrical converters of the AWS device in order to regulate the power quality and smooth the output power via a DC/DC converter [114]. In the case of a wave energy plant with multiple branches of WEC connection, BES is a possible solution to balance the power and phase difference between the instantaneous output powers from each individual WEC to the power grid. Within acceptable system disturbance, experiments have shown that BES is able to smooth the active power of WEC with improvement of power quality.

CONCLUSIONS

As an indispensable stage in the energy conversion process, electrical control strategies are taking a more and more important role and concern in how to harvest energy efficiently from ocean waves. The intermittent nature of ocean waves is a challenge that needs to be met in order to secure a good quality and reliability of power supply.

This paper has categorized the WECs into three groups with regard to the mechanism of the wave energy conversion. Different control topologies of the conversion system utilized in typical examples of WECs are explained. In Table 6, these three categories are listed, and different WECs with related control systems are presented and classified. Control strategies are also listed according to the different WECs discussed in the paper.

Table 6. Three categories of WECs.

Type		Device	Testing location	Rated power	Conversion system	Control strategies	Ref.
Oscillating water column		LIMPET	Islay, Scotland	500 kW	Turbine-generator	(a) Speed control (b) Power control (c) Torque control	[55]
		Pico plant	Azores, Portugal	100 kW	Turbine-generator	(a) Speed control	[60]
				700 kW		(b) Airflow control	[61]
Overtopping		Wave Dragon	Portugal	1.5 MW	Turbine-generator	(a) DPTC-SVM	[66]
						(b) DTC-SVM	[67]
			Nissum Bredning, Denmark	7 MW		(c) APF	[68]
				11 MW			
Attenuator	Hydraulic system	Wave bob	Ireland	500 kW	Hydraulic system	Hydraulic parallel circuit	[70]
							[72]
		Pela	Orkney,	750 k	Hydra	Frequency	[73]

		mis	Scotland	W	ulic system	tuning	[79]
							[80]
Direct drive system		Lyse kil Proje ct	Lysekil, Sweden	10 k W	Linear generat or	Passive rectification DC control Resonant circuit	[86]
							[87]
							[88]
							[89]
		AWS	Portugal	2 M W	Linear generat or	(a) Feedback linearization control (b) Stiffness control	[96]
							[98]
							[102]
							[39]
							[94]
		Oreg on L10	Oregon, USA	10 k W	Linear Genera tor	PADA system	[104]
							[110]
							[111]

One conclusion is that the control strategies are very dependent on the mechanical structure of the WECs. For example, OWCs are mainly focused on rotational speed control of the turbine and airflow control of the valve, while Point Absorbers are more concerned with the electrical control of the WECs. Although different control strategies are of course comparable when applied to a specific WEC device, it is not easy to make parallel comparisons between different WECs even within the same basic category, thus making it difficult to reach a conclusion on which overall control strategy, if any, that is optimal for wave energy extraction. Fundamental understanding of wave energy absorption, see Section 2, tells us what we are trying to achieve with the electrical control, but the necessary mechanical and electrical engineering choices taken in the design of a WEC strongly affects and constrains the possible control strategies. Taking the Wave Bob and the Lysekil Project as an example – in spite of both WEC technologies being based on linear drive point absorbers, the different power take-offs and choice of electrical system affects the possibilities of applying, e.g. a reactive control strategy. Conversely, for some control strategies on overtopping devices and

OWCs, i.e. the controls of the turbine and the generator, there are similarities due to the form and purpose of the electrical conversion stage.

As indicated by this paper, although many studies have been performed on active control, the wave energy community would benefit a lot from more comparative studies and experimental verification of theory. Today the published knowledge of electrical control varies a good deal between the three general concepts and between the specific WEC technologies. Some technologies benefit from long periods of operation, e.g. some OWCs, where different control strategies have been developed and tested over time, while other technologies still await the experiences from experiments in the sea. The authors suggest that the control methods in Table 6 be used as a starting point for future comparative studies. Furthermore, it is of course valuable to study practical experiments of various control strategies carried out in the wave energy test sites, and to compare the experimental results with that from simulation modeling. Moreover, it is expected that much experience and knowledge on control strategies and related energy absorption is unpublished and owned by commercial wave energy developers. Although this information would be much appreciated by the scientific community that it is a lot to ask from industry to disseminate some of their intellectual property.

Although not much research has been presented on energy storage for wave energy as of yet, this often important and integrated part of the energy conversion strategies has been considered briefly in the paper. In wave energy, short time energy storage is often needed for smoothening of the output power and for improving the power quality and the dynamic response. Because of the nature of oceans waves, with large power variations on the second scale, the authors see energy storage as a natural and important future research topic for wave energy.

In general, due to the necessity of large-scale offshore installations and experiments, wave energy research and development is expensive. However, the physical nature of the resource, e.g. the large potential as a source of renewable energy for the world's societies, the relatively high density of energy, and its availability and predictability, stands as clear motivations to the substantial ongoing activity worldwide. Strategies for active control, often in combination with energy storage, have the potential to dramatically affect the absorbed energy and hence the economy of the devices. The complexity of the electrical control strategies and related technologies affect the risk of experiments in terms of survivability and cost, which is something that needs to be taken into consideration by the wave energy developers.

ACKNOWLEDGMENTS

The work is supported by SweGRIDS, KIC InnoEnergy-CIPOWER, Vetenskapsrådet, Statkraft AS, Fortum OY, the Swedish Energy Agency, Seabased Industry AB, Chinese Scholarship Council, Draka Cable AB, the Gothenburg Energy Research Foundation, Falkenberg Energy AB, Helukabel, Proenviro, ÅF Group, Vinnova, the Foundation for the Memory of J.Gust, Richert, the Göran Gustafsson Research Foundation, Vargöns Research Foundation, Swedish Research Council grant no. 621-2009-3417 and the Wallenius Foundation for their support of the project.

REFERENCES

1. Dean RG, Dalrymple RA. Water wave mechanics for engineers and scientists. World Scientific; 1984.
2. Falnes J. Ocean waves and oscillating systems. Cambridge University Press; 2002.
3. Cruz J. Ocean wave energy, current status and future perspectives. Germany: Springer-Verlag; 2008.
4. Eriksson M. Modelling and experimental verification of PhD thesis Uppsala University, Electricity; 2007.
5. Waters R. Energy from ocean waves: full scale experimental verification of a wave energy converter PhD thesis Uppsala University, Electricity; 2008.
6. Charlier RH, Justus JR. Ocean energies: environmental, economic and technological aspects of alternative power sources. Elsevier; 1993.
7. Sabzehgar R, Moallem M. A review of ocean wave energy conversion systems. In: Electrical power & energy conference (EPEC), 2009 IEEE; 2009. p. 1–6.
8. Mohamed KH, Sahoo NC, Ibrahim TB. A survey of technologies used in wave energy conversion systems. In: 2011 International conference on energy, automation, and signal (ICEAS); 2011, p. 1–6.
9. Heath TV. A review of oscillating water columns. Philos Trans R Soc A 2012;370:235–45.
10. Falcão AF. Wave energy utilization: a review of the technologies. Renew Sustain Energy Rev 2010;14:899–918.
11. Masuda Y. An experience of wave power generator through tests and improvement. Int Union Theor Appl Mech 1986:445–52.

12. Bhattacharyya R, McCormick ME. Wave energy conversion. 1st ed. Amsterdam; Boston: Elsevier; 2003.
13. Amundarain M, Alberdi M, Garrido AJ, Garrido I, Maseda J. Wave energy plants: control strategies for avoiding the stalling behaviour in the Wells turbine. Renew Energy 2010;35:2639–48.
14. Pico. OWC wave power plant on the Azores-a late success n.d. Available online ⟨http://wavec.org/client/files/AF_Folheto_Maio_ING.pdf⟩.
15. Wave Energy Centre. Pico OWC plant, 2013. Available online ⟨http://www.pico-owc.net/⟩.
16. Jasinski M, Malinowski M, Kazmierkowski MP, Sorensen HC, Friis-Madsen E, Swierczynski D. Control of AC/DC/AC converter for multi MW wave dragon offshore energy conversion system. In: IEEE international symposium on industrial electronics; 2007. p. 2685–90.
17. Tapered Channel Wave Energy. Tapchan Model; 2010. Available online ⟨http://taperedchannelwaveenergy.weebly.com/index.html⟩.
18. Thorpe T. An overview of wave energy technologies: status, performance and costs n. d. wave power: moving towards commercial viability. Westminster, London: Broadway House; 1999.
19. Kofoed JP, Frigaard P, Friis-Madsen E, Sørensen HC. Prototype testing of the wave energy converter wave dragon. Renew Energy 2006;31:181–9.
20. Wave Dragon. Wave Dragon has started the development of a 1.5 MW North Sea Demonstrator n.d. Available online ⟨http://www.wavedragon.net/index.php?option=com_content&task=view&id=42&Itemid=67⟩.
21. Count BM, Miyazaki T. Hydrodynamic studies on floating attenuator wave energy devices. Central electricity generating board technology planning and research division; 1983.
22. Gardner FE. Learning experience of AWS pilot plant test offshore Portugal. In: Proceedings of 6th European wave energy conference; 2005.
23. Beirdol P, Valério D, da Costa JS. Linear model identification of the Archimedes wave swing. In: International conference on power engineering, energy and electrical drives, vol. 12; 2007. p. 660–665.
24. Lejerskog E, Gravråkmo H, Savin A, Strömstedt E, Tyrberg S, Haikonen K, et al. Lysekil research site, Sweden: a status update. In: 9th European Wave and Tidal Energy Conference. Southampton, UK; 2011.
25. Seabased AB. Wave energy converter; 2013. Available online ⟨http://www.seabased.com/⟩.
26. Von Jouanne A, Brekken T. Wave energy research, development and demonstration at Oregon State University. Power and energy society general meeting, vol. 24; 2011. p. 1–7.

27. Kazmierkowski MP, Jasiń ski M. Power electronics for renewable sea wave energy. In: 12th international conference on optimization of electrical and electronic equipment (OPTIM); 2010. p. 4–9.

28. Weber J. Wavebob – research, and development network and tools in the context of systems engineering. In: Proceedings of the 8th European wave and tidal energy, Uppsala, Sweden; 2009.

29. Clément A, McCullen P, Falcão A, Fiorentino A, Gardner F, Hammarlund K, et al. Wave energy in Europe: current status and perspectives. Renew Sustain Energy Rev 2002;6:405–31.

30. Budar K, Falnes J. A resonant point absorber of ocean-wave power. Nature 1975;256:478–9.

31. Yemm R, Pizer D, Retzler C, Henderson R. Pelamis: experience from concept to connection. Philos Trans R Soc A 2012;370:365–80.

32. Pelamis Wave Power Ltd. The Pelamis absorbs the energy of ocean waves and converts it into clean, green electricity; 2013. Available online ⟨http://www.pelamiswave.com/image-library⟩.

33. Khan J, Bhuyan G, Moshref A, Morison K, Pease JH, Gurney J. Ocean wave and tidal current conversion technologies and their interaction with electrical networks. In: Power and energy society general meeting-conversion and delivery of electrical energy in the 21st century, vol. 20; 2008. p. 1–8.

34. Alberdi M, Amundarain M, Garrido AJ, Garrido I, Casquero O, De la Sen M. Complementary control of oscillating water column-based wave energy conversion plants to improve the instantaneous power output. IEEE Trans Energy Convers 2011;26:1021–32.

35. Takao M, Setoguchi T, Kaneko TH, Takao M, Maeda T, Inoue M. Impulse turbine for wave power conversion with air flow rectification system. Int J Offshore Polar Eng 2002;12(2).

36. Narayanan V, Murthy BK, Bose S, Sridhara Rao G. Dynamic analysis of a gridconnected induction generator driven by a wave-energy turbine. Power electronics. In: Proceedings of the 1996 international conference on drives and energy systems for industrial growth; 8 January 1996. p. 433–8.

37. Falcão AF, Vieira LC, Justino PAP, Andre JMCS. By-pass air-valve control of an OWC wave power plant. J Offshore Mech Arctic Eng 2003;125:205–10.

38. Setoguchi T, Takao M. Current status of self rectifying air turbines for wave energy conversion. Energy Convers Manag 2006;47:2382–96.

39. Wu F, Zhang XP, Ju P, Sterling MJH. Optimal control for AWS-based wave energy conversion system. IEEE Trans Power Syst 2009;24:1747–55.

40. Alberdi M, Amundarain M, Garrido AJ, Garrido I, Maseda FJ. Fault-ridethrough capability of oscillating-water-column-based wave-powergeneration plants equipped with doubly fed induction generator and airflow control. IEEE Trans Ind Electron 2011;58:1501–17.

41. Justino PAP, Falcão AF. Rotational speed control of an OWC wave power plant. J Offshore Mech Arctic Eng 1999;121:65–70.
42. Justino PAP, Falcão AF. Control simulation of an OWC wave power plant. In: Second European wave power conference; 1995. p. 268–72.
43. Falcão AF de O, Justino PAP. OWC Wave Energy converters with valveconstrained air flow. In: Second European wave power conference; 1995. p. 187–194.
44. Sarmento A, Brito-Melo A, Neumann F. Results from sea trials in the OWC European wave energy plant at Pico, Azores. Invited paper for WREC-IX Florence, Italy; 2006.
45. Rao SS, Murthy BK. Control of Induction generator in a Wells turbine based wave energy system. In: International conference on power electronics and drives systems (PEDS), vol. 2; 2005. p. 1590–4.
46. Munoz-Garcia A, Lipo TA, Novotny DW. A new induction motor V/f control method capable of high-performance regulation at low speeds. IEEE Trans Ind Appl 1998;34:813–21.
47. Falcão AF, Justino PAP. OWC wave energy devices with air flow control. Ocean Eng 1999;26:1275–95.
48. Gato LMC, Warfield V, Thakker A. Performance of a high-solidity wells turbine for an OWC wave power plant. J Energy ResourTechnol 1996;118: 263–268.
49. Sarmento A, Falcão AF. Wave generation by an oscillating surface-pressure and its application in wave energy extraction. J Fluid Mech 1985;150:467–85.
50. Sarmento A, Melo A, Pontes MT. The influence of the wave climate on the design and annual production of electricity by OWC wave power plants. J Offshore Mech Arctic Eng 2003;125:139–44.
51. Boake CB, Whittaker TJ, Folley M, Ellen H. Overview and initial operational experience of the LIMPET wave energy plant. In: Proceedings of the twelfth international offshore and polar engineering conference; 2002. p. 586–94.
52. Voith Hydro Wavegen Ltd. Islay Wave Cam; 2013. Available online ⟨http://www.wavegen.co.uk/what_we_offer_limpet_islay_wavecam.htm⟩.
53. Heath T, Whittaker TJT, Boake CB. The design, construction and operation of the LIMPET wave energy converter (Islay, Scotland). In: 4th European wave energy conference; 2000. p. 49–55.
54. Rashid H. Power electronics: circuits, devices, and applications. Pearson/Prentice Hall; 2004.
55. Wilfred PJ. Control system for wave energy devices. U.S. Patent GB2424042 (A); 2006.
56. Falcão AF, Sabino M, Whittaker T, Lewis A. Design of a shoreline wave power pilot plant for the island of Pico, Azores. In: Proceedings of the 2nd European wave power conference; 1995. p. 87–93.

57. Neumann F, Izan IL. Pico OWC-the Frog Prince of Wave Energy? Recent autonomous operational experience and plans for an open real-sea test centre in semi-controlled environment. In: Proceedings of the 9th European wave and tidal energy conference (EWTEC 2011), Southampton, UK; 2011.

58. Salter SH, Taylor JRM, Caldwell NJ. Power conversion mechanisms for wave energy. Proc Inst Mech Eng M: J Eng Maritime Environ 2002; 216:1–27.

59. Le Crom I, Melo A, Neumann F, Sarmento A. Numerical estimation of incident wave parameters based on the air pressure measurements in Pico OWC Plant. In: Proceeding of the 8th European wave and tidal energy conference (EWTEC), Uppsala; 2009.

60. Falcão AF, Rodrigues RJA. Stochastic modelling of OWC wave power plant performance. Appl Ocean Res 2002;24:59–71.

61. Falcão AF. Control of an oscillating-water-column wave power plant for maximum energy production. Appl Ocean Res 2002;24:73–82.

62. Duckers L. Wave power. Eng Sci Educ J 2000;9:113–22. Y. Hong et al. / Renewable and Sustainable Energy Reviews 31 (2014) 329–342 341

63. Frigaard P. Overtopping measurements on the Wave Dragon Nissum Bredning prototype. In: The proceedings of the fourteenth (2004) international offshore and polar engineering conference; 2004. p. 210–216.

64. Polinder H, Scuotto M. Wave energy converters and their impact on power systems. In: International conference on future power systems, vol. 18; 2005. pp. 618–623.

65. Jasinski M, Kazmierkowski MP. Direct power and torque control of AC/DC/AC converter with induction generator for renewable energy system–Wave Dragon MW. In: Proceedings of the 6th WSEAS/IASME, Spain; 2006. p. 328–33.

66. Kazmierkowski MP. Control in power electronics. Academic Press; 2002.

67. Zhou Z, Knapp W, MacEnri J, Sorensen HC, Friis Madsen E, Masters I, et al. Permanent magnet generator control and electrical system configuration for Wave Dragon MW wave energy take-off system. IndElectron 2008:1580–5.

68. Rodriguez JR, Dixon LW, Espinoza JR, Pontt J, Lezana P. PWM regenerative rectifiers: state of the art. IEEE Trans Ind Electron 2005;52:5–22.

69. Ricci P, Lopez J, Santos M, Ruiz-Minguela P, Villate JL, Salcedo F, et al. Control strategies for a wave energy converter connected to a hydraulic power takeoff. IET Renew Power Gener 2011;5:234–44.

70. Schlemmer K, Fuchshumer F, Böhmer N, Costello R, Villegas C. Design and control of a hydraulic power take-off for an axis-symmetric heaving point absorber. In: Proceedings of the NINTH European wave and tidal energy conference, Southampton; 2011.

71. Babarit A, Clément AH. Optimal latching control of a wave energy device in regular and irregular waves. Appl Ocean Res 2006;28:77–91.
72. Villegas C. Implementation of a pitch stability control for a wave energy converter. Wavebob Ltd.; n.d.
73. Cândido JJ, Justino PAP. Frequency, stochastic and time domain models for an articulated wave power device. Proc ASME 2008.
74. Falnes J, Lillebekken PM. Budal's latching-controlled-buoy type wave-power plant. In: 5th European wave energy conference; 2003, p. 17–20.
75. Lopes MFP, Hals J, Gomes RPF, Moan T, Gato LMC, Falcão AF. Experimental and numerical investigation of non-predictive phase-control strategies for a point-absorbing wave energy converter. Ocean Eng 2009;36:386–402.
76. Carcas MC. The OPD Pelamis WEC: current status and onward programme (2002). Int J Ambient Energy 2003;24:21–8.
77. Pizer DJ, Retzler CH, Yemm RW. The OPD Pelamis: experimental and numerical results from the hydrodynamic work program. In: Proceedings of Fourth European wave energy conference, Aalborg; 2000. p. 227–33.
78. Henderson R. Design, simulation, and testing of a novel hydraulic power take-off system for the Pelamis wave energy converter. Renew Energy 2006;31:271–83.
79. Retzler C, Pizer D, Henderson R, Ahlqvist J, Cowieson F, Shaw M, et al., Recent advances in the numerical and experimental modelling programme. In: Proceeding on 6th European wave energy conference, vol. 5; 2003. p. 59–66.
80. Retzler C. Measurements of the slow drift dynamics of a model Pelamis wave energy converter. Renew Energy 2006;31:257–69.
81. Stansell P, Pizer DJ. Maximum wave-power absorption by attenuating line absorbers under volume constraints. Appl Ocean Res 2013;40:83–93.
82. Leijon M, Bernhoff H, Agren O, Isberg J, Sundberg J, Berg M, et al. Multiphysics simulation of wave energy to electric energy conversion by permanent magnet linear generator. IEEE Trans Energy Convers 2005;20:219–24.
83. Krishna R, Kottayil SK, Leijon M. Direct predictive current control of grid connected neutral point clamped inverter for wave power extraction. In: 2010 international symposium on power electronics electrical drives automation and motion (SPEEDAM); 2010. p. 175–9.
84. Uppsala Universitet. Lysekil Project; 2013. Available online 〈http://www.el.angstrom.uu.se/forskningsprojekt/WavePower/Lysekilsprojektet.html〉.
85. Waters R, Stalberg M, Danielsson O, Svensson O, Gustafsson S, Stromstedt E, et al. Experimental results from sea trials of an offshore wave energy system. Appl Phys Lett 2007;90 (034105–3).

86. Kurupath V, Ekström R, Leijon M. Optimal constant DC link voltage operation of a wave energy converter. Energies 2013;6(4):1993–2006.

87. Bostrom C, et al. Design proposal of electrical system for linear generator wave power plants. In: Industrial electronics, 2009. IECON'09. 35th Annual conference of IEEE. IEEE; 2009. p. 4393–8.

88. Ekström R, Baudoin A, Rahm M, Leijon M. Marine substation design for gridconnection of a research wave power plant on the Swedish West coast. In: Proceedings of the 10th European wave and tidal conference (EWTEC), Aalborg, Denmark; 2–5 September 2013.

89. Ekergård B, Bostrom C, Hagnestål A, Waters R, Leijon M. Experimental results from a linear wave power generator connected to a resonance circuit. Wiley Interdisciplinary Rev: Energy Environ 2013;2:456–64.

90. Bostrom C, Ekergard B, Leijon M. Electric resonance-rectifier circuit for renewable energy conversion. Appl Phys Lett 2012;100:043511.

91. Bostrom C, Ekergard B, Waters R, Eriksson M, Leijon M. Linear generator connected to a resonance-rectifier circuit. IEEE J Ocean Eng 2013;38:255–62.

92. Da Costa JS, Pinto P, Sarmento A, Gardner F. Modeling of an ocean waves power device AWS. In: Proceedings of 2003 IEEE conference on control applications; 2003. p. 618–623.

93. Polinder H, Damen MEC, Gardner F. Linear PM Generator system for wave energy conversion in the AWS. IEEE Trans Energy Convers 2004;19:583–9.

94. Wu F, Zhang XP, Ju P, Sterling M. Modeling and control of AWS-Based wave energy conversion system integrated into power grid. IEEE Trans Power Syst 2008;23:1196–204.

95. Prado M, Polinder H. Direct drive in wave energy conversion—AWS full scale prototype case study. Power Energy Soc General Meeting 2011:1–7.

96. Valério D, Beirao P, da Costa JS. Feedback linearisation control applied to the Archimedes Wave Swing. In: Mediterranean conference on control and automation; 2007. p. 1–6.

97. Slotine JJE, Li W. Applied nonlinear control. New Jersey: Prentice hall; 1991.

98. Valério D, Beirao P, Costa JS. Reactive control and phase and amplitude control applied to the Archimedes Wave Swing. In: 17th international offshore (ocean) and polar engineering conference and exhibition, Lisbon; 2007.

99. Babarit A, Duclos G, Clément AH. Comparison of latching control strategies for a heaving wave energy device in random sea. Appl Ocean Res 2004;26:227–38.

100. Falnes J. Optimum control of oscillation of wave-energy converters. Int J Offshore Polar Eng 2002:12.

101. Greenhow M, White SP. Optimal heave motion of some axisymmetric wave energy devices in sinusoidal waves. Appl Ocean Res 1997;19:141–59.

102. Valério D, Beirão P, Costa JS. Optimisation of wave energy extraction with the Archimedes Wave Swing. Ocean Eng 2007;34:2330–44.

103. Brooking PRM, Mueller MA. Power conditioning of the output from a linear vernier hybrid permanent magnet generator for use in direct drive wave energy converters. IEEE Proc – Gener Transm Distrib 2005;152:673–81.

104. Brekken TKA, von Jouanne A, Han HY. Ocean wave energy overview and research at Oregon State University. In: Power electronics and machines in wind applications, vol. 24; 2009. p. 1–7.

105. Amon E, Brekken TKA, von Jouanne A. A power analysis and data acquisition system for ocean wave energy device testing. Renew Energy 2011;36(7): 1922–1930.

106. Mohan N. Advanced electric drives – analysis, modeling and control using Simulink. Minneapolis, MN: Minnesota Power Electronics Research & Education; 2001.

107. Bostrom C, Waters R, Lejerskog E, Svensson O, Stalberg M, Stromstedt E, et al. Study of a wave energy converter connected to a nonlinear load. IEEE J Ocean Eng 2009;34:123–7.

108. Luan H, Onar OC, Khaligh A. Dynamic modeling and optimum load control of a PM Linear Generator for ocean wave energy harvesting application. In: Applied power electronics conference and exposition; 2009. p. 739–43.

109. Henshaw NR. A force control algorithm for a wave energy linear test bed Master Thesis Oregon State University; 2009.

110. VanderMeulen AH. Novel control of a permanent magnet linear generator for ocean wave energy applications Master Thesis Oregon State University; 2007.

111. Amon EA, Schacher AA, Brekken TKA. A novel maximum power point tracking algorithm for ocean wave energy devices. Energy Convers Congress Expos 2009:2635–41.

112. Muthukumar S, Kakumanu S, Sriram S, Jayashankar V. Energy storage considerations for a stand-alone wave energy plant. In: IEEE international conference on electric machines and drives, vol. 15; 2005. p. 193–8.

113. Shek JKH, Macpherson DE, Mueller MA. Power conversion for wave energy applications. Power Electron Mach Drives 2010:1–6.

114. Wu F, Zhang XP, Ju P. Application of the battery energy storage in wave energy conversion system. Sustainable power generation and supply. In: International Conference on SUPERGEN' 09; 2009. p. 1–4.

CITATION

Yue Hong, Rafael Waters, Cecilia Boström, Mikael Eriksson, Jens Engström, Mats Leijon, Review on electrical control strategies for wave energy converting systems, Renewable and Sustainable Energy Reviews, Volume 31, March 2014, Pages 329-342, ISSN 1364-0321, http://dx.doi.org/10.1016/j.rser.2013.11.053.

CHAPTER 4

An Optimal PR Control Strategy with Load Current Observer for a Three-Phase Voltage Source Inverter

Xiaobo Dou [1], Kang Yang [1], Xiangjun Quan [1], Qinran Hu [1], Zaijun Wu [1], Bo Zhao [2], Peng Li [2], Shizhan Zhang [1] and Yang Jiao [1]

[1]Department of Electrical Engineering, Southeast University, No. 2 Sipailou, Nanjing 210096, China;
[2]State Grid Zhejiang Electric Power Research Institute, Hangzhou 310014, China;

ABSTRACT

Inverter voltage control is an important task in the operation of a DC/AC microgrid system. To improve the inverter voltage control dynamics, traditional approaches attempt to measure and feedforward the load current, which, however, needs remote measurement with communications in a microgrid system with distributed loads. In this paper, a load current observer (LCO) based control strategy, which does not need remote measurement, is proposed for sinusoidal signals tracking control of a three-phase inverter of the microgrid. With LCO, the load current is estimated precisely, acting as the feedforward of the dual-loop control, which can effectively enlarge the stability margin of the control system and improve the dynamic response to load disturbance. Furthermore, multiple PR regulators are applied in this strategy conducted in a stationary αβ frame to suppress the transient fluctuations and the total harmonic distortion (THD) of the output voltage and achieve faster transient performance compared with traditional dual-loop control in a rotating dq0 frame under instantaneous change of various types of load (*i.e.*, balanced load, unbalanced load, and nonlinear load). The parameters of multiple PR regulators are analyzed and selected through the root locus method and the stability of the whole control system is evaluated and analyzed. Finally, the validity of the proposed approach is verified through simulations and a three-phase prototype test system with a TMS320F28335 DSP.

INTRODUCTION

Microgrid is commonly defined as an integrated power system consisting of distributed generators (DGs), distributed energy storages (DSs), and interconnected load, which can operate in grid-connected mode or in intentional islanded mode [1]. To increase the stability of the system, making it more immune to perturbations, such as changes in the loading conditions or changes in the electrical energy production due to environmental variability, the employment of energy storage devices such as a battery have become a solution [2,3].

In the microgrid converter system, power converters operate in different modes, the control techniques of the inverter, which are described in references [4,5], are different from each other. In general, in the grid-connected mode, the control target of the inverter is current or direct power, while in the islanded mode, it is the voltage that the inverter controls, which is analogous to an uninterruptible power supply (UPS) for its local loads [3,6]. This paper will focus on the study of the voltage control of a three-phase inverter.

Various voltage control strategies of a three-phase inverter in the microgrid have been researched. The open-loop control strategy is used in many occasions due to its easy control and good dynamic response [7]. However, the steady-state load voltage may not be compensated to the desired value owing to voltage drop across the filter and the line resistance. To avoid the drawbacks of the open-loop control strategy, a load voltage feedback strategy is adopted in [8]. However, such a single feedback scheme is still inadequate for the bad load regulating characteristic and the poor dynamic response. In [9,10,11], a double feedback control scheme is employed with an inner current loop within an outer voltage loop. The inner current loop can be formed using either the filter inductor current or the filter capacitor current. However, the low frequency disturbance from the load will affect the load voltage performance significantly. In order to improve the performance of the dual-loop control strategy, load current is added into the internal current loop in this paper to enhance the dynamic response and enlarge the stability margin of the control system. Hence, precise information of the load current is required to implement the internal current loop.

Since the load current is under low frequency, a current sensor can be used to measure load current [12,13]. However, these sensors suffer from several practical difficulties. Firstly, a correction algorithm has to be added to compensate for the time-varying bias caused by current sensors, which will increase the complexity of the control system in terms of implementation. Secondly, the use of current sensors requires a very precise and noise-free differential amplifier, which is difficult to realize in an actual experimental system. Furthermore, the use of current sensors is not flexible for the change of terminal load and these sensors are very costly, which contributes to the overall cost of the inverter.

Considering all the fundamental issues regarding the current sensors mentioned above, current senseless techniques are very advantageous in this particular application. In this paper, an observer is proposed to estimate the load current. This observer not only can provide a clean and noise-free estimation of the load current but can also provide a good cost-free solution for this situation. Furthermore, the control strategy of inverters in this paper is based on stationary frame (α-β coordinates). Compared with the control strategy under rotating frame (d-q coordinates) using PARK transformation, the input control signal under α-β coordinates does not need to be decoupled, which makes the control strategy easier to design [14,15,16,17]. However in reference [17], the sinusoidal values make it impossible for traditional PI regulators to track accurately without static error. Furthermore, a series of low frequency harmonics is introduced due to the dead-time of one bridge, a PI regulator cannot eliminate this kind of harmonics. To solve these problems, the proportional resonant (PR) is introduced in the control strategy [18,19]. The static error of signal control at specific frequency points can be totally eliminated because the ideal PR regulator can offer infinite gain at these points. Moreover, compared with a conventional tandem double loop PR control, the proposed control strategy has high precision and a physical property with an extensible and zero loosely coupled characteristic.

As a result, the proposed control strategy chosen in this paper introduces a load current observer and adopts multiple R regulators under α-β coordinates to track the sinusoidal voltage values. It introduces the load current as the feedforward of the internal current loop, which accelerates the regulating rate of the internal current loop, improves the dynamic response of the system effectively, and restrains the influence on the system caused by the load fluctuation. Also, it has an outstanding performance in voltage control, such as fast transient response, few steady-state errors, and harmonic rejection for low THD under various types of loads (*i.e.*, balanced load, unbalanced load and nonlinear load) for the use of multiple R regulators. To confirm the commonality and feasibility of the proposed control approach, simulations and experiments were performed through the Matlab/Simulink software and through an experimental platform with a TMS320F28335 DSP.

The rest of this paper is arranged as follows. Section 2 presents the basic mathematical model of the three-phase inverter, PR regulator and load current observer. Section 3 details the structure of the proposed control strategy and the design parameters of multiple PR regulators. Section 4 shows the results of simulation and the laboratory test of the control strategy. Section 5 concludes the paper and remarks on possible future research.

MATHEMATICAL MODEL

Model of Three-Phase Inverter

The model of a three-phase inverter can be divided into two categories: one category is based on a synchronous frame system (dq-frame) that usually adopts a PI regulator; another category is built upon a stationary frame system (αβ-frame) that employs the proportional-resonant (PR) regulator.

Figure 1 shows a typical three-phase voltage-source inverter with a RLC filter. The behavior of the circuit can be modeled as follows by applying Kirchhoff's law:

$$\mathbf{v_{abc}} - \mathbf{u_{abc}} = L\frac{di_{abc}}{dt} + Ri_{abc} \tag{1}$$

$$C\frac{du_{abc}}{dt} = \mathbf{i_{abc}} - \mathbf{i_{labc}} \tag{2}$$

where i_{abc}, u_{abc}, and i_{labc} denote the vectors of the three-phase inductor current, capacitor voltage, and load current, respectively.

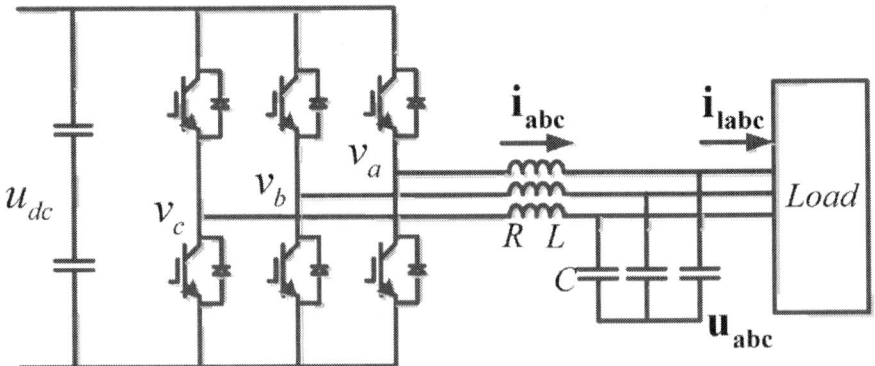

Figure 1. Diagram of a three-phase inverter with a RLC output filter.

The aforementioned state Equations (1) and (2) in the stationary abc reference frame can be transformed to the equations in the stationary $\alpha\beta$ frame by the Clark Transform:

$$T_{3s \to 2s} = \frac{2}{3} \begin{bmatrix} 1 & -\frac{1}{2} & -\frac{1}{2} \\ 0 & \frac{\sqrt{3}}{2} & -\frac{\sqrt{3}}{2} \end{bmatrix} \tag{3}$$

The equation of voltage and current in the stationary $\alpha\beta$ frame can be listed as follows:

$$\mathbf{v}_{\alpha\beta} - \mathbf{u}_{\alpha\beta} = L\frac{d\mathbf{i}_{\alpha\beta}}{dt} + R\mathbf{i}_{\alpha\beta} \tag{4}$$

$$C\frac{d\mathbf{u}_{\alpha\beta}}{dt} = \mathbf{i}_{\alpha\beta} - \mathbf{i}_{l\alpha\beta} \tag{5}$$

where,

$$\mathbf{i}_{\alpha\beta} = \begin{bmatrix} i_\alpha & i_\beta \end{bmatrix}^{\mathrm{T}}, \quad \mathbf{u}_{\alpha\beta} = \begin{bmatrix} u_\alpha & u_\beta \end{bmatrix}^{\mathrm{T}}, \quad \mathbf{i}_{l\alpha\beta} = \begin{bmatrix} i_{l\alpha} & i_{l\beta} \end{bmatrix}^{\mathrm{T}},$$
$$\mathbf{v}_{\alpha\beta} = \begin{bmatrix} v_\alpha & v_\beta \end{bmatrix}^{\mathrm{T}}.$$

In consideration of the inexistence of zero sequence components, the expression for zero sequence is removed.

On account of the full symmetry of the α axis and β axis, the analysis and control design can be handled on the α axis independently.

Consequently, the model expressed by Equations (4) and (5) can be transformed to the format of the state space:

$$
\begin{bmatrix} \dot{U}_\alpha \\ \dot{U}_\beta \\ \dot{i}_\alpha \\ \dot{i}_\beta \end{bmatrix} = \begin{bmatrix} 0 & 0 & \frac{1}{C} & 0 \\ 0 & 0 & 0 & \frac{1}{C} \\ -\frac{1}{L} & 0 & -\frac{R}{L} & 0 \\ 0 & -\frac{1}{L} & 0 & -\frac{R}{L} \end{bmatrix} \begin{bmatrix} U_\alpha \\ U_\beta \\ i_\alpha \\ i_\beta \end{bmatrix} + \begin{bmatrix} 0 & 0 & -\frac{1}{C} & 0 \\ 0 & 0 & 0 & -\frac{1}{C} \\ \frac{1}{L} & 0 & 0 & 0 \\ 0 & \frac{1}{L} & 0 & 0 \end{bmatrix} \begin{bmatrix} V_\alpha \\ V_\beta \\ i_{l\alpha} \\ i_{l\beta} \end{bmatrix} \quad (6)
$$

The control model of the proposed three-phase inverter with an LC output filter in the stationary αβ-frame is illustrated in Figure 2.

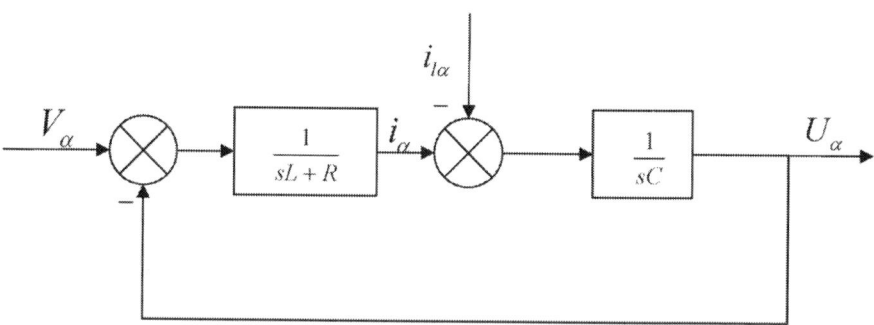

Figure 2. Block diagram of control structure of the three-phase inverter.

As can be seen in Equation (6) and Figure 2, the state variables in the α axis have no coupling relation with those in the β axis. Hence, the balanced three-phase PWM inverter can be equal to two independent single-phase PWM inverters after transformation from the abc reference frame to the stationary αβ frame, which indicates that the control strategy of single-phase inverter can also be implemented under the stationary αβ frame. In this way, the design of the controller can be simplified mathematically.

Model of PR Regulator

According to the internal model principle [20], the best solution for tracking the sinusoidal reference signal and harmonic rejection is the adoption of the PR regulator. An ideal PR regulator has the form as follows:

$$G_{PR}(s) = K_p + \frac{2K_i s}{s^2 + w_h{}^2} \qquad (7)$$

where and K_p and K_i are the proportional and the resonant control gain respectively, wh is the fundamental angular frequency of the source. Figure 3 shows the Bode plot of $G_{PR}(s)$, when Kp = 1, Ki = 1 and wh = 314 rad/s. It can be seen from Figure 3 that the infinite gain of the ideal transfer function is at the frequency of wh, is infinite, and there is no phase shift or gain at other frequencies.

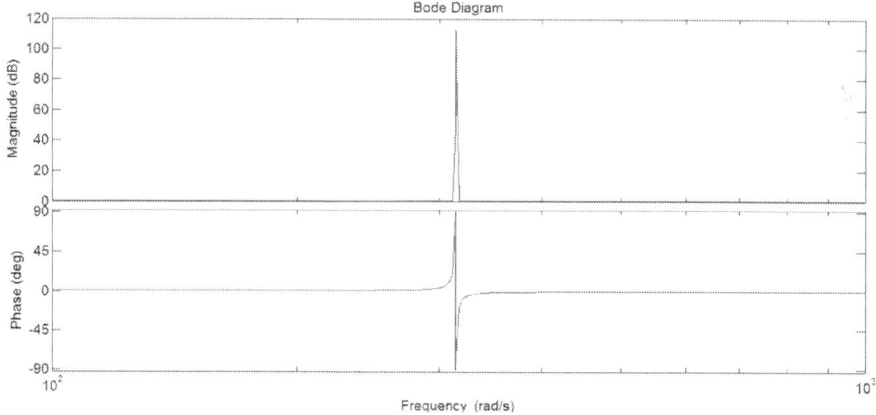

Figure 3. Bode diagram of ideal proportional-resonant (PR) regulator.

Model of Load Current Observer (LCO)

Feedforward control of the load current can provide excellent dynamics in three-phase PWM inverter control. However, the situation becomes more complicated and difficult in a common AC bus, feeding multiple inverter devices, in which multiple current sensors and communication channels are required. To obtain the required AC link load current for feedforward signal without adding additional current sensors or communication channels, a feedforward load current is obtained from a second-order current observer.

In the AC link:

$$C\frac{dUc}{dt} = I_0 - I_{ld} \qquad (8)$$

where Uc is the voltage of the parallel capacitor, I_0 and I_{ld} are the currents of inductor and load respectively.

The discretization of Equation (8) can be expressed as:

$$\begin{cases} \dot{X} = AX + Bu \\ \quad y = CX \end{cases} \tag{9}$$

where:

$$A = \begin{bmatrix} 0 & -\frac{1}{C} \\ 0 & 0 \end{bmatrix}, \; B = \begin{bmatrix} \frac{1}{C} \\ 0 \end{bmatrix}, \; C = [1 \;\; 0], \; X = \begin{bmatrix} Uc \\ I_{ld} \end{bmatrix}, \; \dot{X} = \begin{bmatrix} \dot{Uc} \\ \dot{I}_{ld} \end{bmatrix},$$

$$y = Uc, \, u = I_0.$$

The dynamic equation of state observer based on the Luenberger method can be expressed as:

$$\begin{cases} \dot{\hat{X}} = A\hat{X} + Bu - H(\hat{y} - y) \\ \quad \dot{\hat{y}} = C\hat{X} \end{cases} \tag{10}$$

where $\hat{X}$ and $\hat{y}$ are the observer state vector and output respectively. H=[h1 h2]$^\mathrm{T}$ is the observer gain vector.

The block diagram of the state observer derived from Equation (10) is shown in Figure 4.

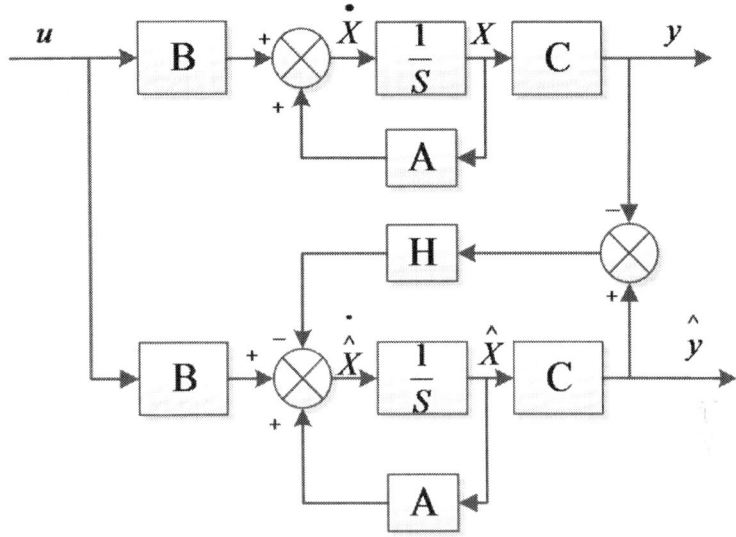

Figure 4. Block diagram of the state observer.

From Equations (9) and (10):

$$\dot{\hat{X}} - \dot{X} = (A - HC)(\hat{X} - X) \qquad (11)$$

The state vector depends only on the initial error and is independent of the input Uc for convergence to a zero state. In order that the vector difference Equation (11) is asymptotically stable, the characteristic equation:

$$\det(sI - (A - HC)) = 0 \qquad (12)$$

must have all its roots within the unit circle, *i.e.*, $|zi|<1$, $i = 1,2,...,$m. It is known that if the matrix $[A-HC]$ yields a stable solution, the error vector will converge to zero from any initial error. Hence, stability is achieved by selecting the observer feedback gain matrix$H$, so that the roots of the characteristic equation have negative real parts. By expanding and discretizing Equation (10), the observation equation of the load current can be expressed as:

$$\begin{cases} \dot{\hat{Uc}} = \frac{1}{C}\left[I_0 - \hat{I}_{ld}\right] - h_1\left[\hat{Uc} - Uc\right] \\ \\ \dot{\hat{I}}_{ld} = -h_2\left[\hat{Uc} - Uc\right] \end{cases} \qquad (15)$$

From the above equation, an alternative expression of LCO in a continuous-time domain is illustrated in Figure 5.

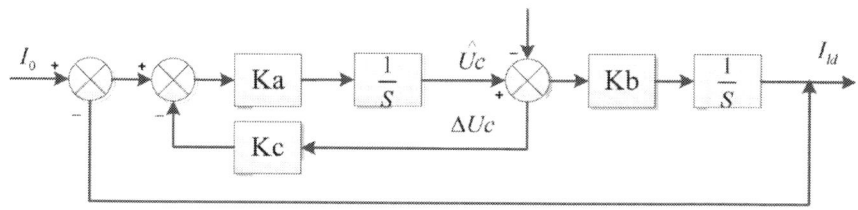

Figure 5. Block diagram of load current observer (LCO).

where, Ka=1C/, Kb=h2, Kc=Ch1.

Figure 5 clearly shows the physical dynamic process of LCO. The value of C used in the observer can be chosen the same as the physical capacitance of the AC link.

The design of the observer needs to confirm the observer gain H=[h1 h2]$^\mathrm{T}$ so that the roots of the characteristic equation have negative real parts. For the observer being a second order system, the desired characteristic equation is assumed as

$$\det[\lambda I - (A-HC)]=0$$

$$\det\left[\lambda I - (A - HC)\right] = 0 \tag{14}$$

$$\begin{vmatrix} \lambda + h_1 & 1/C \\ h_2 & \lambda \end{vmatrix} = \lambda^2 + \lambda h_1 - \frac{h_2}{C} = 0 \tag{15}$$

It is assumed that both of the roots of the characteristic equation are $\lambda 1 = \lambda 2 = k$, (k<0), then

$$\lambda^2 + h_1 \lambda - \frac{h_2}{C} = \lambda^2 - 2k\lambda + k^2 \tag{16}$$

By solving Equation (16), $H = [2k - CK^2]$.

To illustrate the effectiveness of the designed LCO, the response of the three-phase PWM inverters was simulated with the load in AC link changing. Using the parameters given above, with the capacitance set at 15 uF and a sampling frequency of 12,800 Hz, the simulated performance of the observer is illustrated in Figure 6.

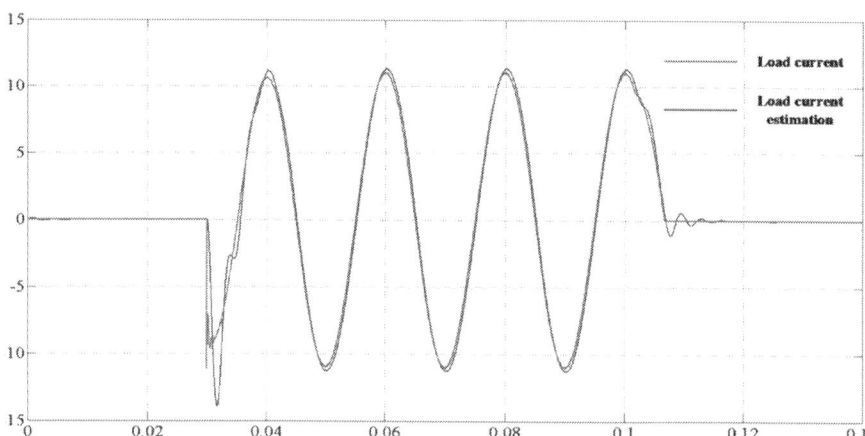

Figure 6. Typical performance of the designed observer.

Since the simulation system is developed to match the actual system very closely, the control parameters can be applied in the experimental voltage control loop. As can be seen in Figure 5, the observer response is still fast enough to estimate the ac link load current with good accuracy and minimum delay.

CONTROL STRUCTURE AND PARAMETERS DESIGN

Based on the inverter diagram as shown in Figure 1, the system control strategy under the stationary frame is illustrated in Figure 7.

As shown in Figure 7, the control strategy used in this paper is presented. The output capacitor voltage is regarded as the control target, which is controlled under the stationary frame. The voltage of capacitors is controlled by multiple PR regulators, the output of which is the reference value of the internal current loop. The internal current loop is control by P regulators, after which, the pulse-width-modulation (PWM) module is adopted to generate the drive of IGBT. These five PR regulators are used to mitigate the harmonics with different orders and all of these PR regulators show no effect on each other. Hence, the multiple PR regulators can be considered separately in the system control structure or parameters design, which can make it easier and more effective to analyze the control structure and design control parameters.

The structure of traditional dual-loop control of the inverter, which is based on rotating frame (dq0 coordinates) is shown in Figure 8a. In this control strategy, PI regulators are adopted to track static signals and the input control signal under dq0 coordinates needs to be decoupled. Furthermore, a series of low frequency harmonics is introduced due to the dead-time of one bridge; the PI regulator cannot eliminate this kind of harmonics.

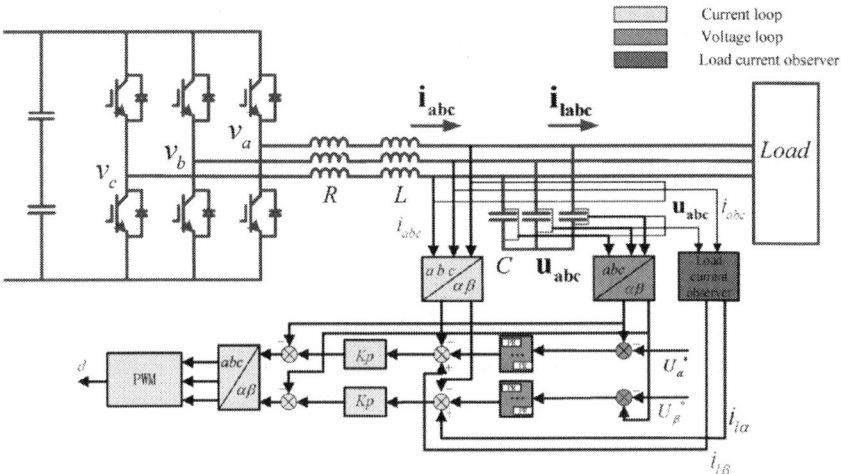

Figure 7. Inverter control system under stationary frame.

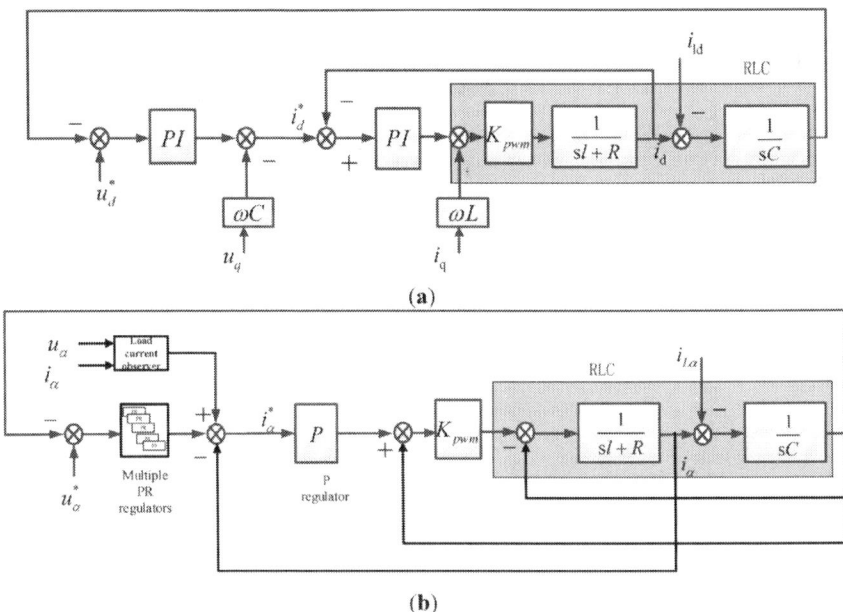

Figure 8. Structure of controllers. (a) Traditional inverter controller; (b) The proposed inverter controller with LCO.

In this paper, the controller of inverter is designed as in Figure 8b. Compared with the traditional control strategy, the input control signal under α-β coordinates needs not to be decoupled, which makes the control strategy easier to design. As for aspects of zero steady-state error and anti-interference ability, the PR regulator shows superior performance to the PI regulator. What is more, the PR regulator reduces the harmonic content of the output AC voltage and improves the system performance [21].

As shown in Figure 8b, the output of the LCO, of which the input is capacitor voltage and inductor current in the AC link, acts as the feedforward of the internal current loop of the proposed controller. Considering the independence of every PR regulator of the multiple PR regulators module in mitigating harmonics of output signals, the closed-loop and open-loop transfer functions of the control system, which includes only one PR regulator that regulates the fundamental harmonic, can be deduced as:

$$G_{cl}(S) = \frac{a_0 s^2 + a_1 s + a_2}{b_0 s^4 + b_1 s^3 + (b_2 + a_0)s^2 + (b_3 + a_1) s + (b_4 + a_2)}$$

$$G_{op}(S) = \frac{a_0 s^2 + a_1 s + a_2}{b_0 s^4 + b_1 s^3 + b_2 s^2 + b_3 s + b_4}$$

Where

$a_0 = K_p K_{pwm} P$, $a_1 = 2K_i K_{pwm} P$, $a_2 = K_p K_{pwm} P w^2$, $b_0 = CL$,
$b_1 = CR + CK_{pwm} P$, $\qquad\qquad b_2 = CLw^2 + 1 - K_{pwm}$,
$b_3 = w^2 CR + w^2 CK_{pwm} P$, $b_4 = w^2 (1 - K_{pwm})$.

The open loop transfer function indicates that it is a four-order system and it is complex to choose the proper controller parameters. In order to select the proper parameters, *i.e.*, Kpand Ki, parameter related root locus of the closed-loop transfer function is adopted in the analysis. By changing one of the parameters each time, the influence of every parameter on the system performance is analyzed [18]. The closed-loop root locus diagrams are shown as follows.

Examples of root locus where Kp or Ki1 varies, are illustrated in Figure 9 and Figure 10respectively. As shown in Figure 9, the poles of the closed-loop transfer function vary with the value of Kp while Ki1 = 150 and *p* = 10 and the dynamic performance of the control strategy keeps stable when Kp > 0.0582. Similarly, Figure 10 shows the poles of the closed-loop transfer function varies with the value of Ki1 where Kp = 0.3 and *p* = 10 and the dynamic performance of the control strategy stays stable when Ki1 < 785. It is worth mentioning that the designs of other parameters of 5-multiple PR regulators based on root locus are similar to the given examples.

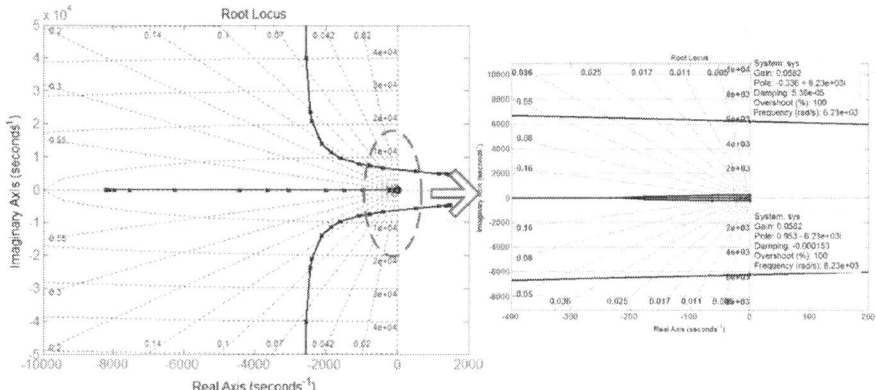

Figure 9. Diagram of root locus with varying parameter Kp while other parameters are maintained constant.

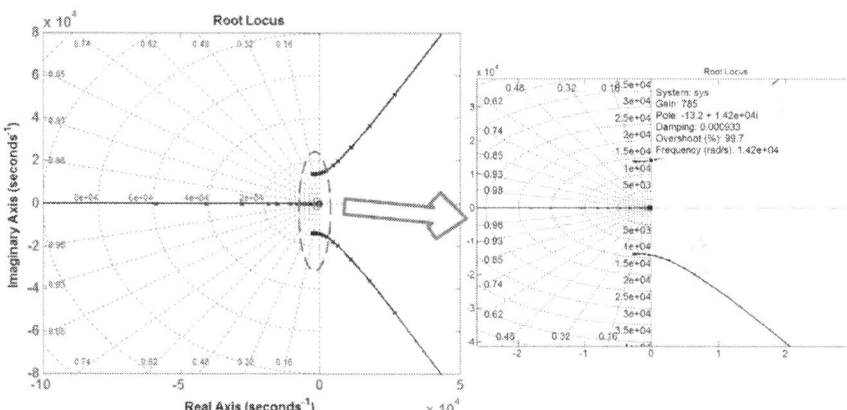

Figure 10. Diagram of root locus with varying parameter Ki1 while other parameters are maintained constant.

Based on the root locus method, the proper parameters of 5-multiple PR regulator are shown in Table 1 and parameters of simulation are shown in Table 2.

Table 1. Parameters of controller.

Parameters	Value
K_p	0.3
K_{i1}	150
K_{i2}	100
K_{i3}	20
K_{i4}	80
K_{i5}	5
P	15

Table 2. Parameters of simulation.

Parameters	Value
DC-link voltage U_{dc}	700 V
Output voltage U_{abc}	220 V
Output frequency f	50 Hz
Switching and Sampling frequency f_s	12,800 Hz
Output Filter R	0.5 Ω
Output Filter L	2 mH
Output Filter C	15 uF

To analyze the stability of the control strategy, the zero-pole distributive chart of the closed-loop transfer function Gop(S) is shown in Figure 11. Based on the analysis and the given parameters above, the location of poles and zeros are $P_1 = -110$, $P_2 = -937$, $P_{3,4} = -2100 \pm i9500$, $Z_1 = -111$, $Z_2 = -889$. The corresponding Bode diagram of the open-loop transfer function Gop(S) is shown in Figure 12.

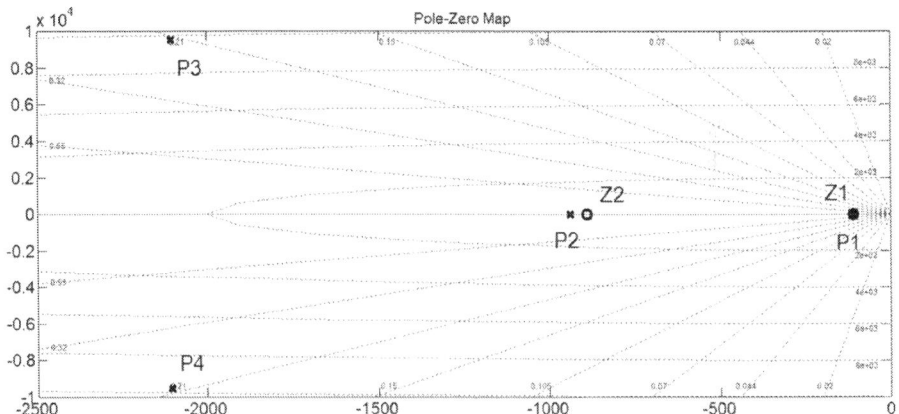

Figure 11. Zero-pole distributive chart of the closed-loop transfer function Gop(S).

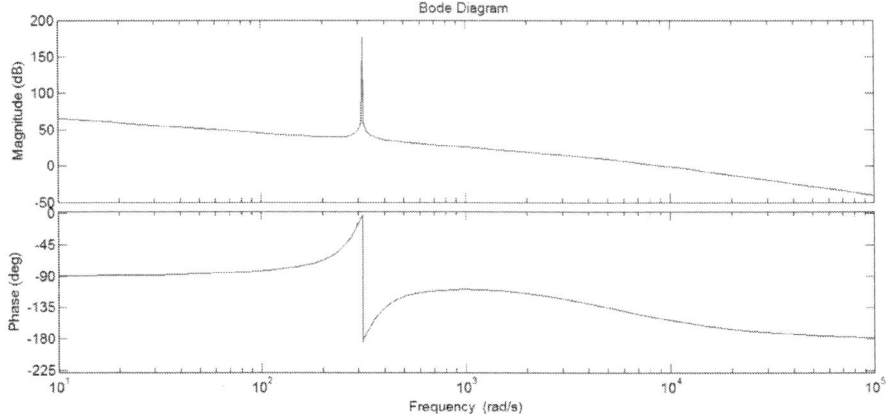

Figure 12. Bode diagram of the open-loop transfer function.

As shown in Figure 12, the close-loop transfer function has a gain greater than 150 dB at a frequency of 341 rad/s, which indicates that the static error of signal control at specific frequency (50 Hz) points can be totally eliminated due to the PR regulator introduced for this control strategy.

SIMULATION AND LABORATORY EXPERIMENTS

Simulation Study

A simulation system was established in the MATLAB/SIMULINK environment to verify the effectiveness of the proposed method for LCO based on the αβ frame applied for the control of the voltage source inverter in the microgrid. To comprehensively understand the advantages of the proposed control strategy and the LCO, several pairs of comparison simulation on the proposed control scheme with and without LCO were conducted. The comparison simulation was performed under the following three conditions:

- Scenario 1: The balanced resistive load is applied to the inverter output terminals at 0.03 s and then cut off at 0.1 s.
- Scenario 2: The unbalanced resistive load is applied to the inverter output terminals at 0.03 s and then cut off at 0.1 s.
- Scenario 3: The nonlinear resistive load is applied to the inverter output terminals at 0.03 s and then cut off at 0.1 s.

Figure 13, Figure 14 and Figure 15 show the simulation results of the proposed voltage controller using MATLAB/Simulink under the three different conditions mentioned above. Comparison simulation on the proposed control with and without LCO were conducted to verify its validity when the load changes.

Under the control of the proposed strategy without LCO, the output voltage had an instantaneous fluctuation of ±60 V in its amplitude when the balanced load was connected to or removed from the inverter output terminals as shown in Figure 13a. Furthermore, it took approximately 0.2 s for the output voltage to return to the steady-state status. While with the proposed LCO-based control strategy, the output voltage variation was mitigated significantly. As shown in Figure 13b, the output voltage variation reduces from 60 V to less than 10 V, which indicates that the proposed strategy with LCO obviously has a better performance in voltage control when the load current provided by the observer is introduced as the feedforward of the control strategy. Furthermore, the transient response of the output voltage waveforms is faster at the moment of the balanced load being instantaneously applied to the inverter output terminal. The advantage of LCO was also verified under Scenario 2, which is displayed in Figure 14a,b. As can be seen in Figure 15, the proposed voltage controller has a good performance under nonlinear load and the THD of the output voltage is kept at 2.95% as shown in Figure 17.

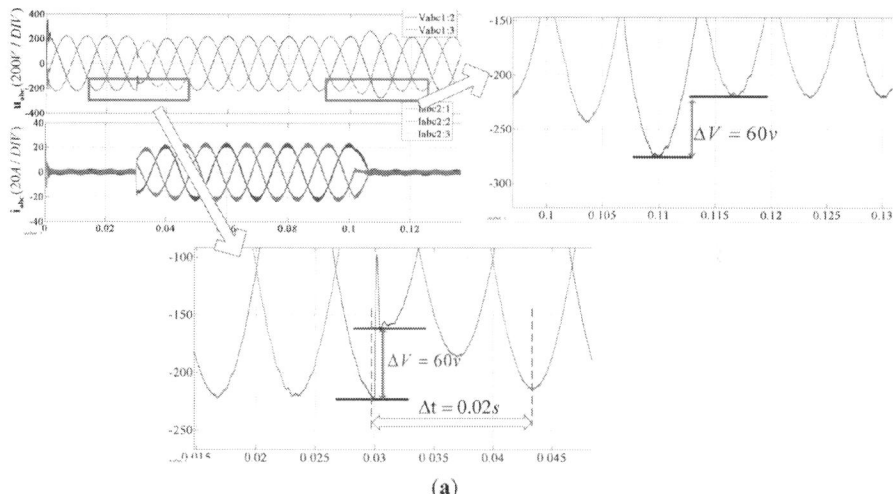

Figure 13. Simulation results of the propose control strategy under Scenario 1: (a) without LCO; (b) with LCO.

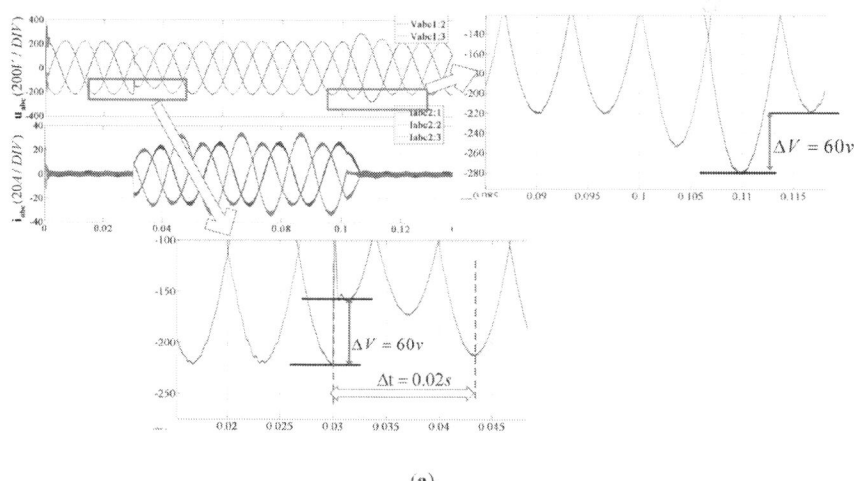

Figure 14. Simulation results of the propose control strategy under Scenario 2: (a) without LCO; (b) with LCO.

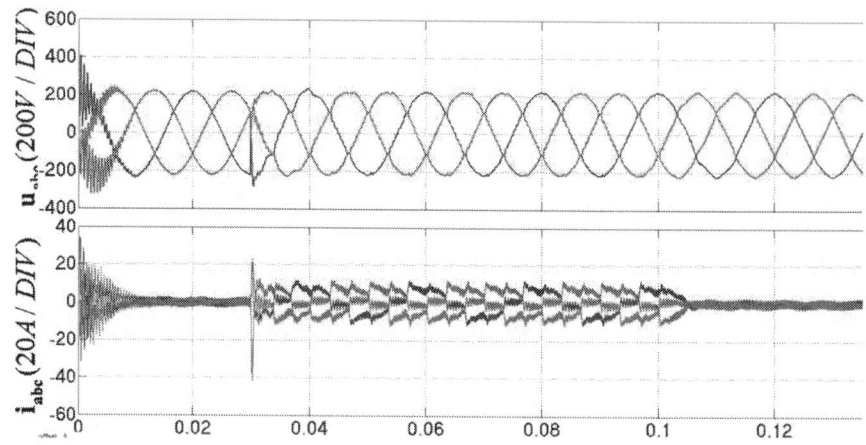

Figure 15. Simulation results of the propose control strategy with LCO under Scenario 3.

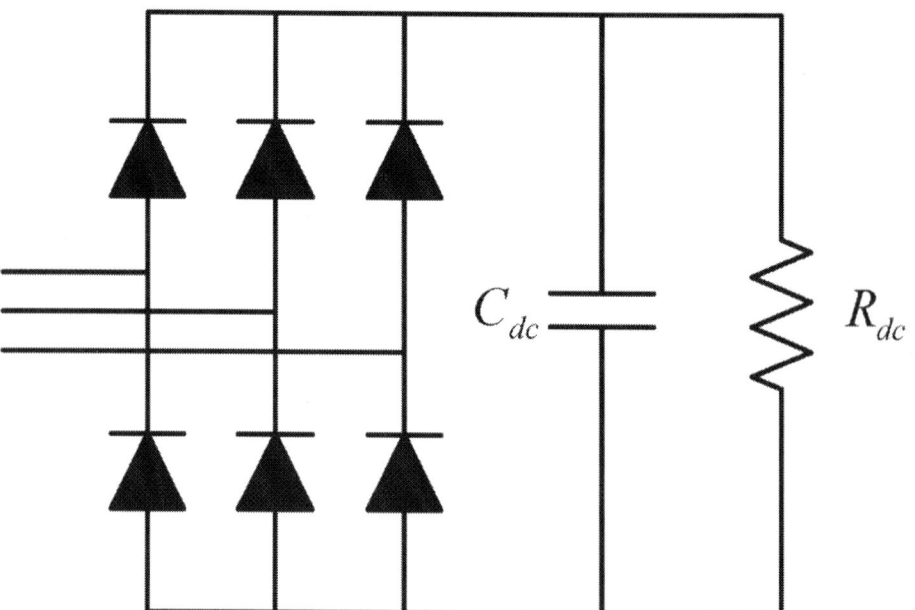

Figure 16. Nonlinear load circuit with a three phase diode rectifier.

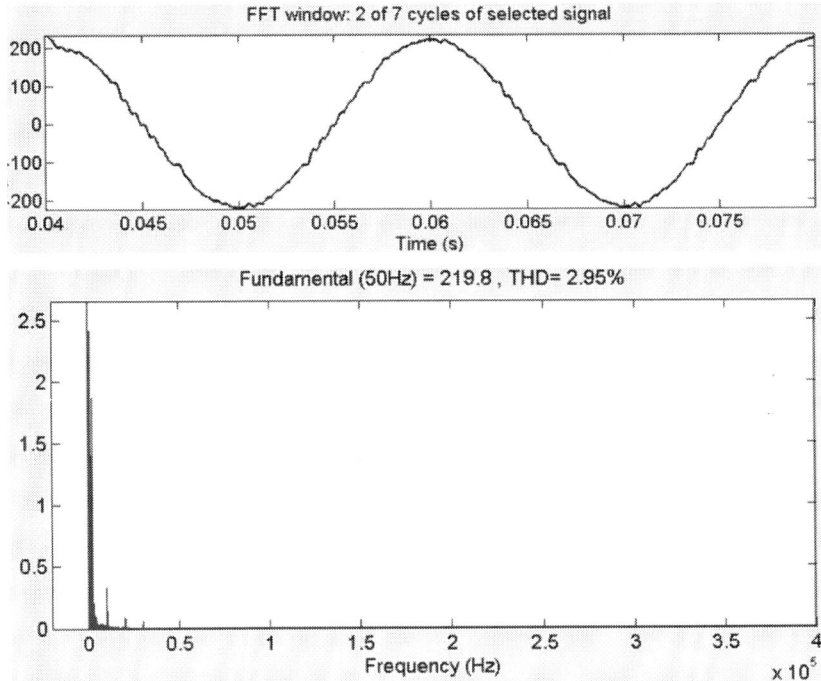

Figure 17. Fast Fourier Transformation (FFT) analysis of output voltage under Scenario 3.

The parameters of the nonlinear load circuit are listed in Table 3.

Table 3. Parameters of the nonlinear load circuit.

Items	Value
R_{dc}	50 Ω
C_{dc}	20 μF

Experimental Test

To verify the effectiveness of the proposed LOC-based control strategy of three-phase inverter under a stationary frame, a three-phase prototype test system as illustrated in Figure 1 was built. The corresponding hardware experiment platform is shown in Figure 18, with its system parameters listed in Table 2.

In the experiments, comparisons of the traditional control and the proposed LOC-based control under balanced load, listed in the simulation part, were also conducted. The experimental results are listed in the following figures.

Figure 19a–c represent the experimental results of the proposed control scheme with and without LCO and traditional dual-loop control scheme when the balanced load is connected to the inverter terminals, respectively. As can be seen in the Figure 19a, the output voltage stays in the steady-state status without any decrease in its amplitude. While in Figure 19b–c, the output voltages both dropped from 311 V to 260 V with nearly 50 V of voltage disturbance at the moment of the load being connected. The experimental results are consistent with the simulation results above, which indicates the control strategy proposed in this paper shows an excellent performance on the dynamic response to the load disturbance. Figure 20 shows a stable state waveform of the system output voltage under unbalanced load, which proves that the proposed control strategy has also an excellent capacity of feeding unbalanced load. Figure 21 presents the waveform of the output voltage while the voltage reference rises to 311 V.

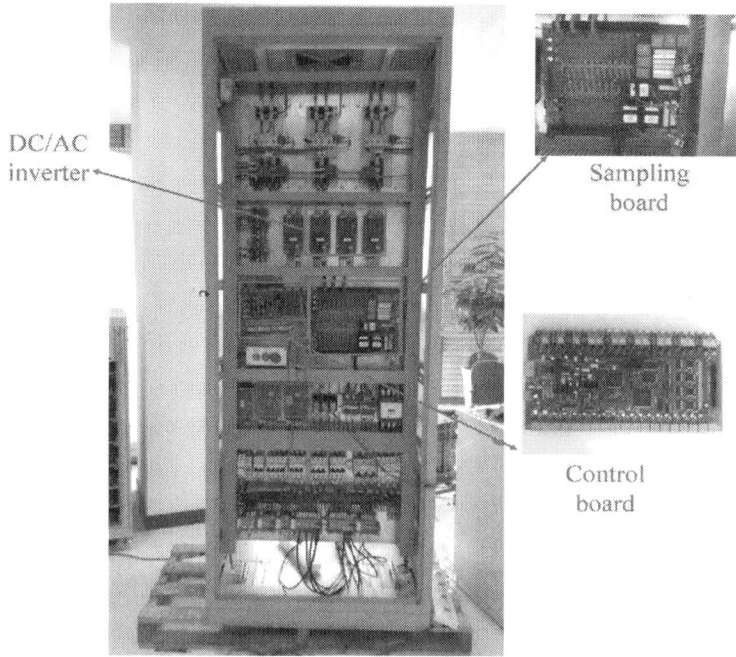

Figure 18. Hardware experiment platform.

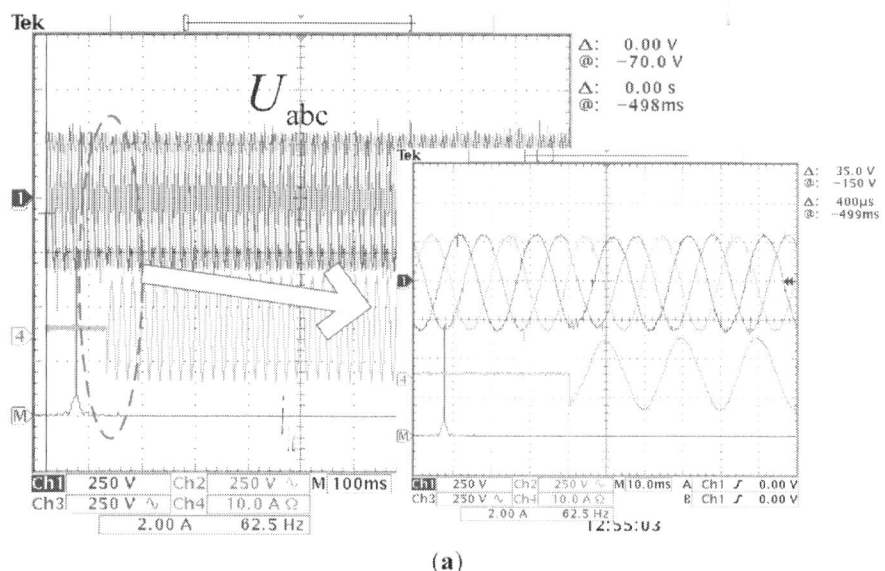

(a)

Figure 19. Experimental results under balanced load switch. (**a**) Proposed control scheme with LCO; (**b**) Proposed control scheme without LCO; (**c**) Traditional dual-loop control scheme.

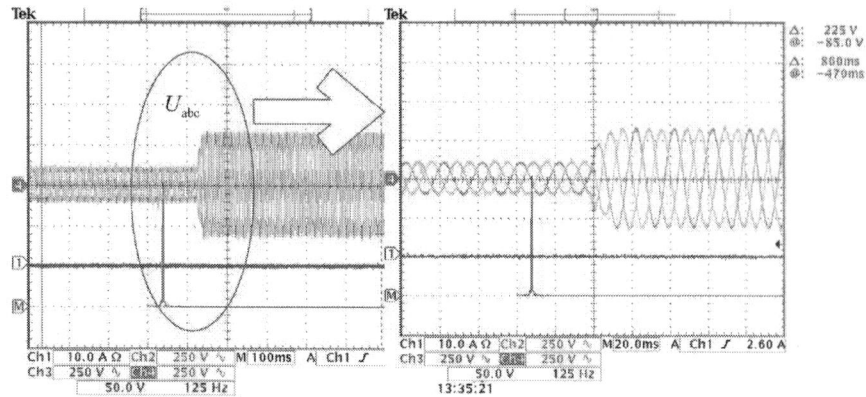

Figure 20. Experimental results of the proposed control strategy with LCO under unbalanced load.

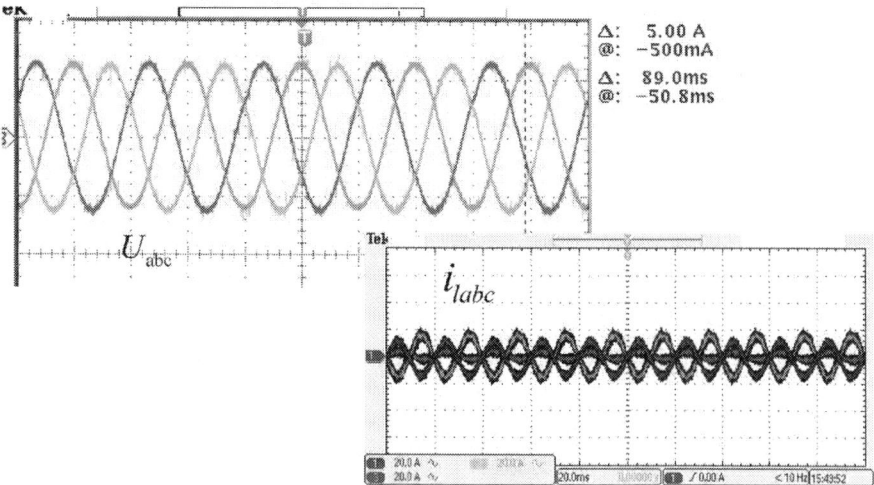

Figure 21. Experimental results of the proposed control strategy with LCO when voltage reference rises to 311 V.

CONCLUSIONS AND FUTURE WORK

This paper has proposed an optimal PR control strategy with LCO for a three-phase inverter of islanded microgrid. The effectiveness and feasibility of the proposed control strategy and its design method were proved by simulations and laboratory experiments.

1. The proposed control strategy is conducted in a stationary $\alpha\beta$ frame, which has a feature of extensibility and suitability and is convenient for the implementation of multiple PR regulators with different center frequencies. Furthermore, it is easy and simple to design parameters of controllers for the reason that control signals under α-β coordinates need do not have to be decoupled.

2. The proposed control strategy applies load current as feedforward of dual-loop control, which makes a great contribution to its outstanding performance in voltage control such as zero steady-state error and low THD. Furthermore, the proposed control strategy has excellent dynamics including quick dynamic response and short regulation time, all of which ensure that the BESS has a strong ability to support unbalanced and nonlinear loads.

3. The proposed strategy can perfectly support renewable energy, which is beneficial for the spread and development of distributed generations.

Meanwhile, further work remains to be done to perfect the strategy. All of the models and analyses in this paper are based on a continuous domain, suitable for high switching frequency situations. The discrete-time model and corresponding controller design should be established and researched for low switching frequency situations.

ACKNOWLEDGMENTS

This research was supported by the National High Technology Research and Development Program of China (Grant No. 2014AA052002), the National Natural Science Foundation of China (Grant No. 51307023 and 51177015), the Program for New Century Excellent Talents in University (Grant No. NCET-13-0129), the Natural Science Foundation of Jiangsu, China (Grant No. BK20130624), and the Technology Research Program of State Grid Corporation of China (SGCC, Grant No. 5211DS150015).

AUTHOR CONTRIBUTIONS

Xiaobo Dou conceived and designed the proposed control strategy and model. Kang Yang significantly contributed to the implementation of the simulation and the writing of the paper. Xiangjun Quan and Yang Jiao contributed mainly to the laboratory test. Zaijun Wu, Qinran Hu, Bo Zhao and Peng Li were responsible for the guidance and a number of key suggestions. Shizhan Zhang contributed mainly to the analysis of the control parameters in this paper, which is of great significance for the proposed control strategy in this paper.

REFERENCES

1. Mehrizi-Sani, A.; Iravani, R. Potential-function based control of a microgrid in islanded and grid-connected modes. *IEEE Trans. Power Syst.* 2010, *25*, 1883–1891.
2. Tan, X.; Li, Q.; Wang, H. Advances and trends of energy storage technology in Microgrid. *Int. J. Electr. Power Energy Syst.* 2013, *44*, 179–191.
3. Serban, I.; Marinescu, C. Control strategy of three-phase battery energy storage systems for frequency support in microgrids and with uninterrupted supply of local loads. *IEEE Trans. Power Electron.* 2014, *29*, 5010–5020.
4. Rocabert, J.; Luna, A.; Blaabjerg, F.; Rodríguez, P. Control of power converters in AC microgrids. *IEEE Trans. Power Electron.* 2012, *27*, 4734–4749.
5. Kazmierkowski, M.P.; Malesani, L. Current control techniques for three-phase voltage-source PWM converters: A survey. *IEEE Trans. Ind. Electron.* 1998, *45*, 691–670.
6. Karimi, H.; Davison, E.J.; Iravani, R. Multivariable servomechanism controller for autonomous operation of a distributed generation unit: Design and performance evaluation. *IEEE Trans. Power Syst.* 2010, *25*, 853–865.
7. Jauch, T.; Kara, A.; Rahmani, M.; Westermann, D. Power quality ensured by dynamic voltage correction. *ABB Rev.* 1998, *4*, 25–36.
8. Perera, A.A.D.R. Development of Controllers for the Dynamic Voltage Restorer. Master's Thesis, Nanyang Technological University, Singapore, 2000.
9. Vilathgamuwa, M.; Perera, A.A.D.R.; Choi, S.S. Performance improvement of the dynamic voltage restorer with closed-loop load voltage and current-mode control. *IEEE Trans. Power Electron.* 2002, *17*, 824–834.

10. Loh, P.C.; Newman, M.J.; Zmood, D.N.; Holmes, D.G. A comparative analysis of multiloop voltage regulation strategies for single and three-phase UPS systems.*IEEE Trans. Power Electron.* 2003, *18*, 1176–1185.

11. Vilathgamuwa, D.M.; Wijekoon, H.M. Control and analysis of a new dynamic voltage restorer circuit topology for mitigating long duration voltage sags. In Proceedings of the 37th Industry Applications Conference (IAS), Pittsburgh, PA, USA, 13–18 October 2002; IEEE: New York, NY, USA, 2002; Volume 2, pp. 1105–1112.

12. Makki, A.; Bose, S.; Giuliante, T.; Walsh, J. Using hall-effect sensors to add digital recording capability to electromechanical relays. In Proceedings of the 63rd Protective Relay Engineers, College Station, TX, USA, 29 March–1 April 2010; IEEE: New York, NY, USA, 2010; pp. 1–12.

13. Pankau, J.; Leggate, D.; Schlegel, D.; Kerkman, R.; Skibiniski, G. High frequency modeling of current sensors. In Proceedings of the 14th Applied Power Electronics Conference and Exposition (APEC'99), Dallas, TX, USA, 1999; IEEE: New York, NY, USA, 1999; Volume 2, pp. 788–794.

14. Hornik, T.; Zhong, Q.C. Control of grid-connected DC-AC converters in distributed generation: Experimental comparison of different schemes. In Proceedings of the Compatibility and Power Electronics (CPE'09), Badajoz, Spain, 5–22 May 2009; IEEE: New York, NY, USA, 2009; pp. 271–278.

15. Yang, S.; Lei, Q.; Peng, F.Z.; Qian, Z. A robust control scheme for grid-connected voltage-source inverters. *IEEE Trans. Ind. Electron.* 2011, *58*, 202–212.

16. Teodorescu, R.; Blaabjerg, F.; Liserre, M.; Loh, P.C. Proportional-resonant controllers and filters for grid-connected voltage-source converters. *IEE Proc.-Electr. Power Appl.* 2006, *153*, 750–762.

17. Ahmed, K.H.; Massoud, A.M.; Finney, S.J.; Williams, B.W. A modified stationary reference frame-based predictive current control with zero steady-state error for LCL coupled inverter-based distributed generation systems. *IEEE Trans. Ind. Electron.*2011, *58*, 1359–1370.

18. Zeng, G.; Rasmussen, T.W. Design of current-controller with PR-regulator for LCL-filter based grid-connected converter. In Proceedings of the 2nd IEEE International Symposium on Power Electronics for Distributed Generation Systems (PEDG), Hefei, China, 16–18 June 2010; IEEE: New York, NY, USA, 2010; pp. 490–494.

19. Shen, G.; Zhu, X.; Zhang, J.; Xu, D. A new feedback method for PR current control of LCL-filter-based grid-connected inverter. *IEEE Trans. Ind. Electron.* 2010, *57*, 2033–2041.

20. Marconi, L.; Teel, A.R. Internal model principle for linear systems with periodic state jumps. *IEEE Trans. Autom. Control* 2013, *58*, 2788–2802.

21. Wu, Y.; Shang, R.; Guo, X.; Li, Y.; Yu, H. The comparative analysis of PI controller with PR controller for the single-phase 4-quadrant rectifier. In Proceedings of the IEEE Transportation Electrification Conference and Expo Asia-Pacific (ITEC Asia-Pacific), Beijing, China, 31 August–3 September 2014; IEEE: New York, NY, USA, 2014; pp. 1–5.

CITATION

Xiaobo Dou, Kang Yang, Xiangjun Quan, Qinran Hu, Zaijun Wu, Bo Zhao, Peng Li, Shizhan Zhang and Yang Jiao, An Optimal PR Control Strategy with Load Current Observer for a Three-Phase Voltage Source Inverter, doi:10.3390/en8087542.

CHAPTER 5

Modeling and Current Programmed Control of a Bidirectional Full Bridge DC-DC Converter

Shahab H. A. Moghaddam, Ahmad Ayatollahi and Abdolreza Rahmati

School of Electrical Engineering, Iran University of Science and Technology, Tehran, Iran

ABSTRACT

Modelling of bidirectional full bridge DC-DC converter as one of the most applicable converters has received significant attention. Mathematical modelling reduces the simulation time in comparison with detailed circuit response; moreover it is convenient for controller design purpose. Due to simple and effective methodology, average state space is the most common method among the modelling methods. In this paper a bidirectional full bridge converter is modelled by average state space and for each mode of operations a controller is designed. Attained mathematical model results are in a close agreement with detailed circuit simulation.

INTRODUCTION

Modeling of DC-DC converter as one of the most applicable industrial converters has aroused a lot of interest. Since modeling gives us information about static and dynamic of the system, it is a crucial factor in design and control. Moreover, attained mathematical model can reduce the simulation time in comparison with the simulation time provided by "cycle by cycle" solving the differential equations of the circuit, as is the case in matlab/ simulink.

With respect to renewable energy systems and optimum use of regenerated energy, interface converters should be capable of transferring power in both directions. So bidirectional dc-dc converters (BDC) are one of the most important interfaces that have applications such as: hybrid or electrical vehicles [1], aerospace systems [2], telecommunications, solar cells, battery chargers [3], DC motor drive circuits [4], uninterruptable power supplies [5-7], etc. so far many BDCs topologies have been introduced and surveyed [8,9]. In applications that transferred power is more than 750 watts, full bridge topology is a proper one [10].

Bidirectional full bridge (FB) converters have been studied in many papers like [11-13]. A general modeling method that develops the discrete time average model is proposed in [12]. The operation period is divided to 3 intervals, the equivalent circuit and the differential equations for each interval are written in matrix form. After solving equations and applying approximation of Taylor expansion, the averaging state vector in half cycle gives us the final answer. Since time domain method employs numerical integration to solve differential equations the analysis is complicated and computationally intensive. Moreover the information about the dependence of the converter's operating conditions on the circuit parameters is not provided [14].

Reference [13] proposes a discrete Small signal model with the amount of considerable calculation, just to predict the peak response of state vectors. There are also some identification-based methods like NARMAX [15, 16] and Hammerstein [17,18] to model the dc-dc converter. Hammerstein model involves of a static nonlinearity followed by a linear discrete-time and time-invariant model, but identification based methods consider the system as a black/gray box, therefore they do not provide any insight into circuit details. So many references use circuit oriented methods to model the converter. Vorperian and Tymerski et al. [19,20] provide the circuit switch model with replacing the PWM switch with its equivalent circuit in order to model the converter. This method may pose some complexity especially in nonbasic topologies. Another circuit oriented method that is introduced by Middlebrook and Cuk [21-23] in 1977 is the average state space. An advantage of the state-space averaging method is its efficiency compared to that of the switched model because there is not any switching frequency ripple and, consequently, the simulation time required by the averaged model is much lower than required by the switched model. Among all methods of modeling, average state space seems to be one of the most common, simplest, and effective methods, so in this paper we model a bidirectional full bridge dc-dc converter with average state space to gain the appropriate transfer functions for controller design purpose in both modes of operation.

PRINCIPLE OF OPERATION

Figure 1 shows the proposed bidirectional full bridge converter where arrows represent the direction of power flow. There are so many configurations for FB converters [11,14,24] with the same basic topology which differs from one another in the case of switching scheme or employing the elements of converter for the purpose of achieving ZVS or ZCS. Since modeling of this converter is the main purpose of this study and Changing modes of operation rely on conditions of application; these conditions are not taken into consideration.

Detailed Circuit description can be reviewed in the literature [24,25]. There are some long and short intervals in each mode, since the short ones are not as significant as long ones they can be left out. Meanwhile Figure 2 shows the basic waveforms and pulse gating of switches in boost mode operation, ignoring the short subintervals, Figure 3 depicts the buck mode waveforms. Small signal model of each mode will be extracted.

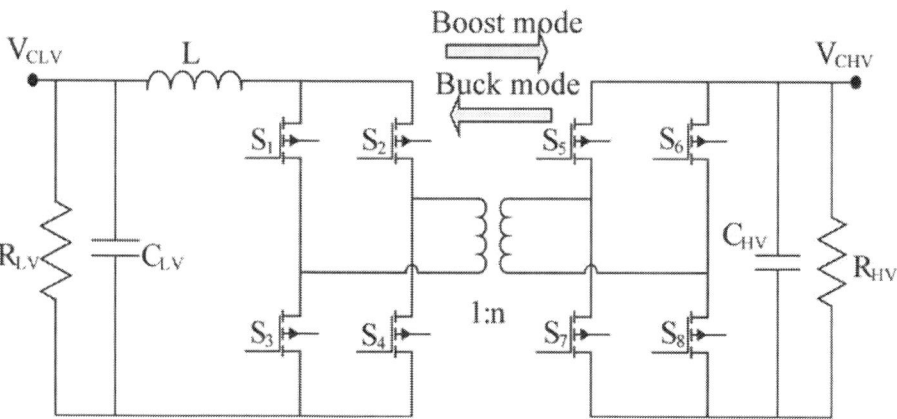

Figure 1. Basic topology of the proposed FB bidirectional converter.

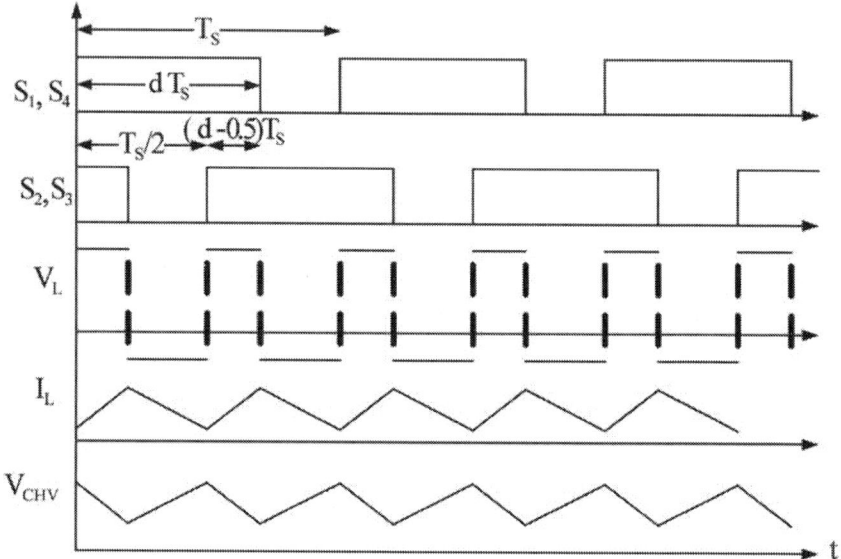

Figure 2. Ideal steady-state voltage and current waveforms of the converter in boost mode operation during one switching cycle.

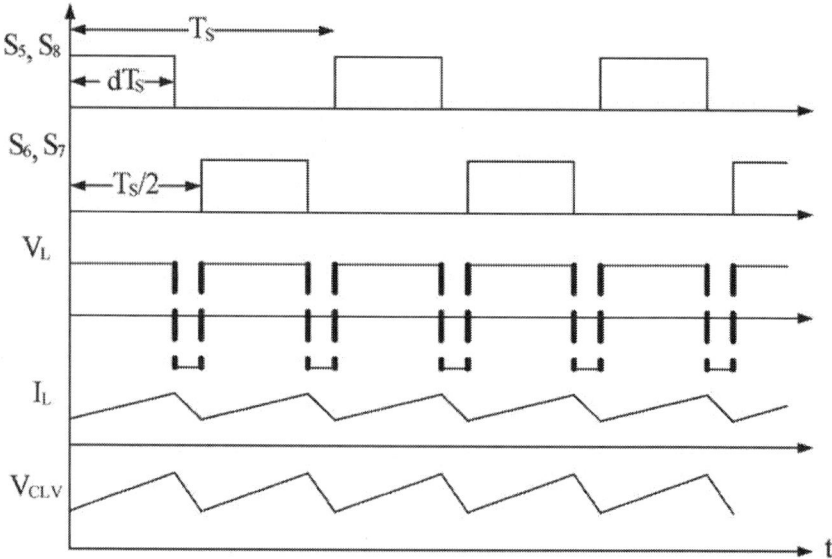

Figure 3. Ideal steady-state voltage and current waveforms of the converter in buck mode operation during one switching cycle.

For simplicity of control, gating circuit, and analysis, it is assume that in every mode only switches of one side are gated meanwhile the opposite ones are operating in their diode modes.

Boost Mode Operation

In boost mode with respect to pulse gating signals, there are two main intervals, consisting of four switches on and two diagonal switches on. When four switches turn on, input inductor voltage is equal to input source (low voltage side) and the inductor current increase proportional to the applied voltage. In this interval inductor saves the energy to transfer it in the next interval. While at the high voltage (HV) side the load is fed by the energy that has been transferred to the output filter (C) during pervious interval.

Next interval usually is known as the energy transfer interval. For instance, assume that S_1 and S_4 are on and S_2 and S_3 are off. The input voltage plus inductor voltage is applied to transformer and is scaled by n factor, ratio of secondary to primary voltage, then will be rectified through the other side of bridge. These processes will be repeated at next half switching cycle with this point that the primary applied voltage in next half switching cycle will be negative but will be rectified in the other side of the bridge.

Buck Mode Operation

According to **Figure 3** that shows the conventional pulse gating, hard switch, for buck mode operation, again there are two main intervals in a half switching cycle, first when diagonal switches turn on, and second one when all switches turn off. When diagonal switches are on, for instance S5 and S8, the power is transferred from high HV side to LV side. In this interval the inductor current increases proportional to the scaled HV side voltage minus output (nominal LV) voltage.

With turning off the switches, next interval starts. Although the existence of leakage inductance prevents the switches go off immediately after applying gate turn off pulses and conduction of switches will continue through parasitic capacitors and diodes, but it is assumed that these subintervals are very short and can be neglected. In off time, the secondary side is only fed by the inductor stored energy, so the inductor current decreases proportional to output voltage. Next half switching cycle is the same, and only applied voltage of HV side is negative that is rectified in LV side.

SMALL SIGNAL MODELING USING AVERAGE STATE SPACE

Employing average state space method is divided into three phases:

1) With respect to switch conditions, the circuit is divided into different subintervals and state equations are written in the matrix form in each interval. State vectors are defined as inductors currents and capacitors voltages.
2) Averaged equations are formed by taking weighted average of state equations of each interval.
3) Averaged equations are written in differential form then linearization is done by perturbing variables. Employing Laplace transform and omitting additional ac and DC terms (only first order ac terms), needed transfer functions are achieved.

For simplicity of modeling the following assumptions can be employed:

• Switches are ideal, there is no parasitic effect in switches;
• Inductor has no resistance;
• Transformer is ideal and there are no leakage and magnetizing inductances;
• Filter capacitors have low ESR (equivalent series resistance) that can be neglected;
• Load is constant and for modeling of the load change an additional current source has been added at the output;
• Each mode (buck or boost modes) starts with zero initial condition.

State equations will be written in each mode separately and with some mathematical operations one can derive needed transfer functions.

Boost Mode State Equations
As we saw in section 2.1 in this mode two main intervals can be assumed. When all four switches conduct the equivalent circuit will be the same as shown in **Figure 4** and the differential equations can be written as follows:

$$v_{IN} = L\frac{di_L}{dt} \implies \frac{di_L}{dt} = \frac{1}{L}v_{IN}$$

(1)

$$\frac{v_c}{R} + C\frac{dv_c}{dt} = i_z \Rightarrow \frac{dv_c}{dt} = \frac{1}{C}i_z - \frac{1}{RC}v_C \tag{2}$$

$$\begin{bmatrix} \dfrac{di_L}{dt} \\ \dfrac{dv_c}{dt} \end{bmatrix} = \begin{bmatrix} 0 & 0 \\ 0 & -\dfrac{1}{RC} \end{bmatrix}\begin{bmatrix} i_L \\ v_C \end{bmatrix} + \begin{bmatrix} \dfrac{1}{L} & 0 \\ 0 & \dfrac{1}{C} \end{bmatrix}\begin{bmatrix} v_{IN} \\ i_z \end{bmatrix} \tag{3}$$

This state lasts for $(d - 0.5)T_s$ where $T_s = 1/f_s$ is period of switching, $d = t_{on}/T_s$ is the effective duty ratio and n is turn ratio of secondary to primary windings. In the next interval when diagonal switches conduct, the equivalent circuit can be sketched as the same in **Figure 5** and the state equations can be written as below:

$$-v_{IN} + L\frac{di_L}{dt} + \frac{v_C}{n} = 0 \Rightarrow \frac{di_L}{dt} = \frac{1}{L}v_{IN} - \frac{1}{nL}v_C \tag{4}$$

$$\frac{i_L}{n} + i_z = \frac{v_C}{R} + c\frac{dv_C}{dt} \Rightarrow \frac{dv_C}{dt} = \frac{1}{nC}i_L + \frac{1}{C}i_z - \frac{1}{RC}v_C \tag{5}$$

$$\begin{bmatrix} \dfrac{di_L}{dt} \\ \dfrac{dv_c}{dt} \end{bmatrix} = \begin{bmatrix} 0 & -\dfrac{1}{nL} \\ \dfrac{1}{nC} & -\dfrac{1}{RC} \end{bmatrix}\begin{bmatrix} i_L \\ v_C \end{bmatrix} + \begin{bmatrix} \dfrac{1}{L} & 0 \\ 0 & \dfrac{1}{C} \end{bmatrix}\begin{bmatrix} v_{IN} \\ i_z \end{bmatrix} \tag{6}$$

$$Y = v_C = v_o \Rightarrow Y = \begin{bmatrix} 0 & 1 \end{bmatrix}\begin{bmatrix} i_L \\ v_C \end{bmatrix} + \begin{bmatrix} 0 & 0 \end{bmatrix}\begin{bmatrix} v_{IN} \\ i_z \end{bmatrix} \tag{7}$$

This state lasts for $(1 - d)T_s$. The output state in both intervals is the same as equation (7). Now averaging the state equations during half switching cycle will result in the equation that has the characteristic of two intervals:

$$A_t = \begin{bmatrix} 0 & \dfrac{2(d-1)}{nL} \\ \dfrac{2(1-d)}{nC} & -\dfrac{1}{RC} \end{bmatrix} \& B_t = \begin{bmatrix} \dfrac{1}{L} & 0 \\ 0 & \dfrac{1}{C} \end{bmatrix} \tag{8}$$

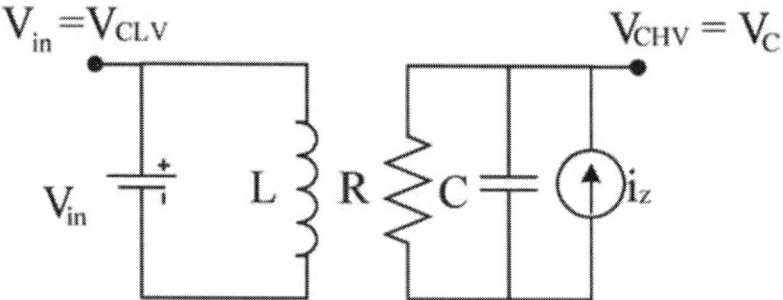

Figure 4. Equivalent circuit of boost mode operation with four LV switches on.

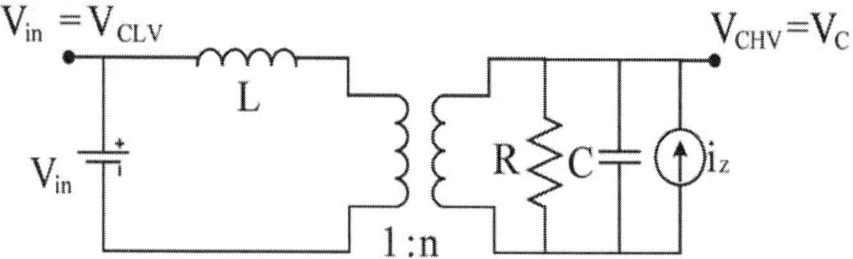

Figure 5. circuit of boost mode operation with two LV switches on.

$$\dot{X} = \begin{bmatrix} \dot{i}_L \\ \dot{v}_C \end{bmatrix} = \begin{bmatrix} 0 & \dfrac{2(d-1)}{nL} \\ \dfrac{2(1-d)}{nC} & -\dfrac{1}{RC} \end{bmatrix} \begin{bmatrix} i_L \\ v_C \end{bmatrix} + \begin{bmatrix} \dfrac{1}{L} & 0 \\ 0 & \dfrac{1}{C} \end{bmatrix} \begin{bmatrix} v_{IN} \\ i_Z \end{bmatrix}$$

(9)

$$Y_c = CX + DU = \begin{bmatrix} 0 & 1 \end{bmatrix} \begin{bmatrix} i_L \\ v_C \end{bmatrix} + \begin{bmatrix} 0 & 0 \end{bmatrix} \begin{bmatrix} v_{IN} \\ i_Z \end{bmatrix}$$

(10)

Perturbing the state equations around their operating points will result in the equations that can be used for deriving the transfer functions; to do so one just need to write voltages and currents in the following forms:

$$\begin{cases} i_L = I_L + \tilde{i}_L \\ V_c = V_C + \tilde{v}_c \\ v_{IN} = V_{IN} + \tilde{v}_{in} \\ i_Z = \tilde{i}_z \\ d = D + \tilde{d} \end{cases}$$

(11)

$$\frac{d\tilde{i}_L}{dt} + \frac{dI_L}{dt} = \frac{2(D + \tilde{d} - 1)}{nL}(V_C + \tilde{v}_c) + \frac{1}{L}(V_{IN} + \tilde{v}_{in})$$

(12)

$$\frac{d\tilde{v}_c}{dt} + \frac{dV_C}{dt} = \frac{2(1 - D - \tilde{d})}{nC}(I_L + \tilde{i}_L)$$
$$- \frac{1}{RC}(V_C + \tilde{v}_c) + \frac{1}{C}\tilde{i}_z$$

(13)

$$\frac{d\tilde{i}_L}{dt} = \frac{2(D + \tilde{d} - 1)}{nL}(V_C + \tilde{v}_c) + \frac{1}{L}(V_{IN} + \tilde{v}_{in})$$

(14)

$$\frac{d\tilde{v}_c}{dt} = \frac{2(1 - D - \tilde{d})}{nC}(I_L + \tilde{i}_L) - \frac{1}{RC}(V_C + \tilde{v}_c) + \frac{1}{C}\tilde{i}_z$$

(15)

For extracting transfer functions additional ac and dc terms can be neglected. With applying Laplace transform to the equations one can achieve the following transfer functions.

$$\frac{\tilde{v}_c}{\tilde{v}_{in}} = \frac{2(1 - D)n}{n^2 CLS^2 + n^2 \dfrac{L}{R}S + 4(1 - D)^2}$$

(16)

$$\frac{\tilde{v}_c}{\tilde{d}} = \frac{-2nLI_L S + 8(1 - D)V_C - 2nV_{IN}}{n^2 CLS^2 + n^2 \dfrac{L}{R}S + 4(1 - D)^2}$$

(17)

$$\frac{\tilde{i}_L}{\tilde{d}} = \frac{2nCV_C S + 8(1 - D)I_L}{n^2 CLS^2 + n^2 \dfrac{L}{R}S + 4(1 - D)^2}$$

(18)

$$\frac{\tilde{v}_c}{\tilde{i}_z} = \frac{n^2 SL}{n^2 CLS^2 + n^2 \dfrac{L}{R}S + 4(1 - D)^2}$$

(19)

Buck Mode State Equations

The method is the same as mentioned in previous section for the boost mode operation. Two main operating intervals are considered to illustrate circuit conditions. No matter what switching scheme is employed, conventional PWM, phase shift or PWM plus phase shift, there is always an effective duty cycle, D_{eff}. Figures 6 and 7 depict the main equivalent circuits meanwhile (20) and (21) describe state equations. Note that in writing equations, output voltage (LVS) is mentioned as V_c and the input voltage at HV side is defined as V_{in}.

$$\begin{bmatrix} \dfrac{di_L}{dt} \\ \dfrac{dv_c}{dt} \end{bmatrix} = \begin{bmatrix} 0 & -\dfrac{1}{L} \\ \dfrac{1}{C} & -\dfrac{1}{RC} \end{bmatrix} \begin{bmatrix} i_L \\ v_C \end{bmatrix} + \begin{bmatrix} \dfrac{1}{nL} & 0 \\ 0 & \dfrac{1}{C} \end{bmatrix} \begin{bmatrix} v_{IN} \\ i_Z \end{bmatrix} \tag{20}$$

This interval will lasts for dT_s seconds. Following interval will last for the remained half cycle, $(0.5 - d)T_s$.

$$\begin{bmatrix} \dfrac{di_L}{dt} \\ \dfrac{dv_c}{dt} \end{bmatrix} = \begin{bmatrix} 0 & -\dfrac{1}{L} \\ -\dfrac{1}{C} & -\dfrac{1}{RC} \end{bmatrix} \begin{bmatrix} i_L \\ v_C \end{bmatrix} + \begin{bmatrix} \dfrac{1}{L} & 0 \\ 0 & \dfrac{1}{C} \end{bmatrix} \begin{bmatrix} v_{IN} \\ i_Z \end{bmatrix} \tag{21}$$

Averaging in half switching cycle and perturbing the voltage and currents will result the needed transfer functions.

$$\frac{\tilde{v}_C}{\tilde{v}_{in}} = \frac{\left(\dfrac{2D}{n}\right)}{CLS^2 + \dfrac{L}{R}S + 1} \tag{22}$$

$$\frac{\tilde{v}_c}{\tilde{d}} = \frac{\left(\dfrac{2V_{IN}}{n}\right)}{CLS^2 + \dfrac{L}{R}S + 1} \tag{23}$$

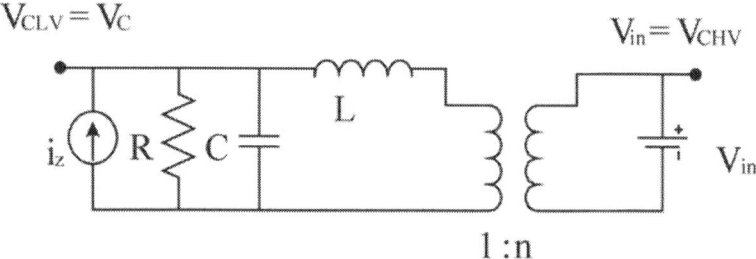

Figure 6. Equivalent circuit of buck mode operation with two HV switches conducting.

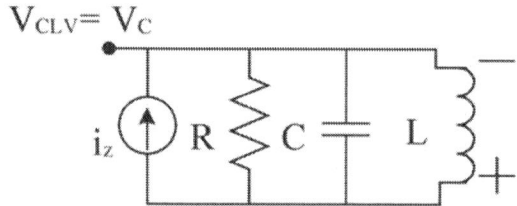

Figure 7. Equivalent circuit of buck mode operation with shorting HV side.

$$\frac{\tilde{i}_L}{\tilde{d}} = \frac{\left(\frac{2CV_{IN}}{n}\right)S + \frac{2V_{IN}}{n}}{CLS^2 + \frac{L}{R}S + 1}$$

(24)

$$\frac{\tilde{v}_c}{\tilde{i}_z} = \frac{LS}{CLS^2 + \frac{L}{R}S + +1}$$

(25)

MODEL VERIFICATION

Having averaged state equation and perturbation one can achieve all transfer functions needed for controller design. In this section verification of the attained averaged model is done by comparing the step responses of transfer functions derived from average state space and simulated circuit in matlab/simulink.

Boost Mode

Start up process of open loop converter system with input voltage of 24 V is simulated under these conditions: $D = 0.6$, $f_s = 20$ kHz, $L = 200$ μH, $C = 50$ μF, $V_{out} = V_c = 300$ V, $P_{in} = P_{out} = 1.5$ KW, $R = 60$ Ω.

The simulated start up process of the predicted mathematical model and detailed circuit simulation with fixed duty cycle is depicted in Figures 8(a) and (b), respectively. Steady state, peak response, and rise time of output voltage in simulink is the same as obtained by the mathematical model but settling time differs 0.00007 seconds which is acceptable.

In order to check control to output and control to inductor current transfer functions, a small step (0.01) can be applied to duty cycle in the both mathematical models and simulated circuit.

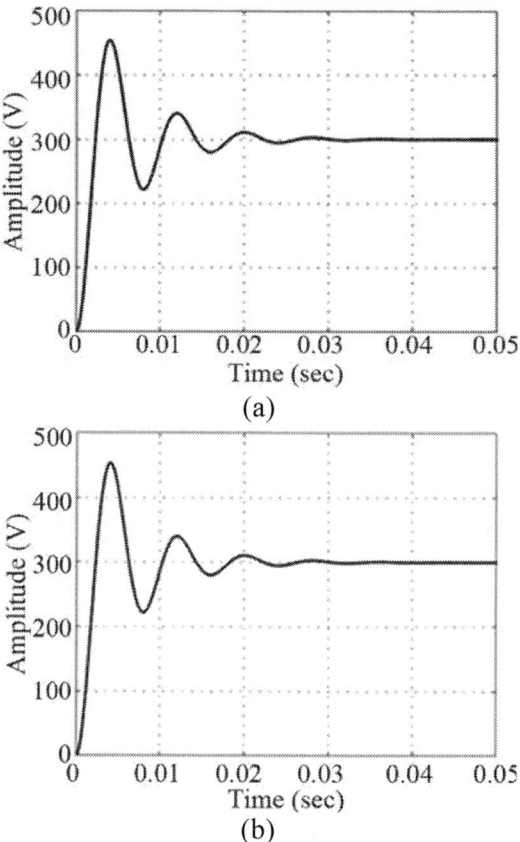

Figure 8. Start-up processes (a) Simulated mathematical averaged; (b) Detailed circuit simulation in matlab/simulink.

Figure 9 shows the output voltage response to duty cycle change and **Figure 10** represents the current waveforms while **Figure 11** represents the output voltage change due to 1A step change in load current that is modeled by a current source, i_z.

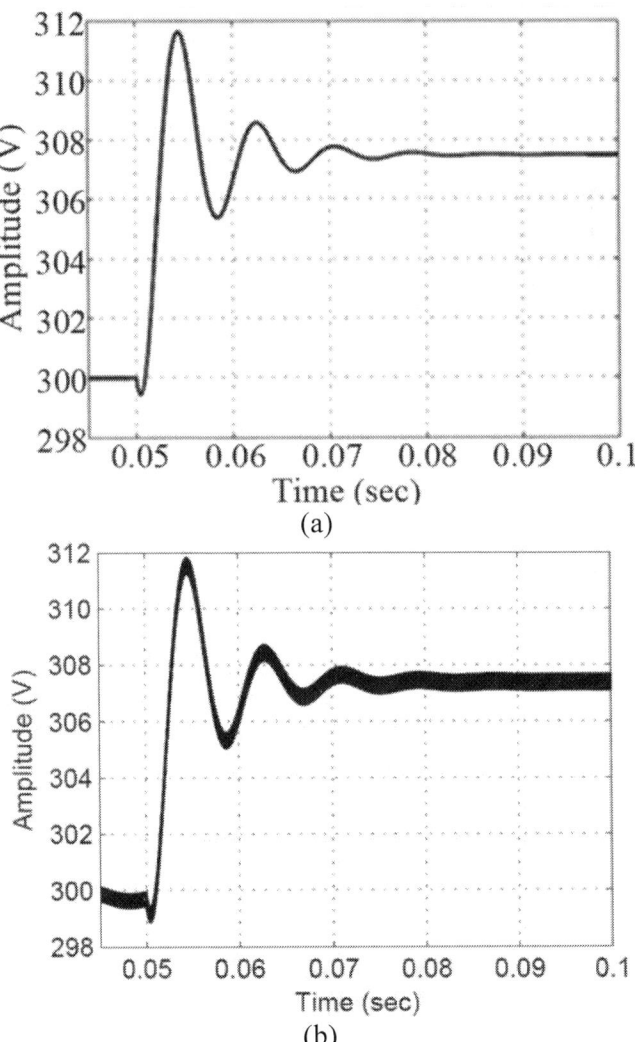

Figure 9. Output voltage in the presence of 0.01 step change in duty cycle. (a) Predicted response by mathematical averaged model; (b) Detailed circuit simulation in matlab/simulink.

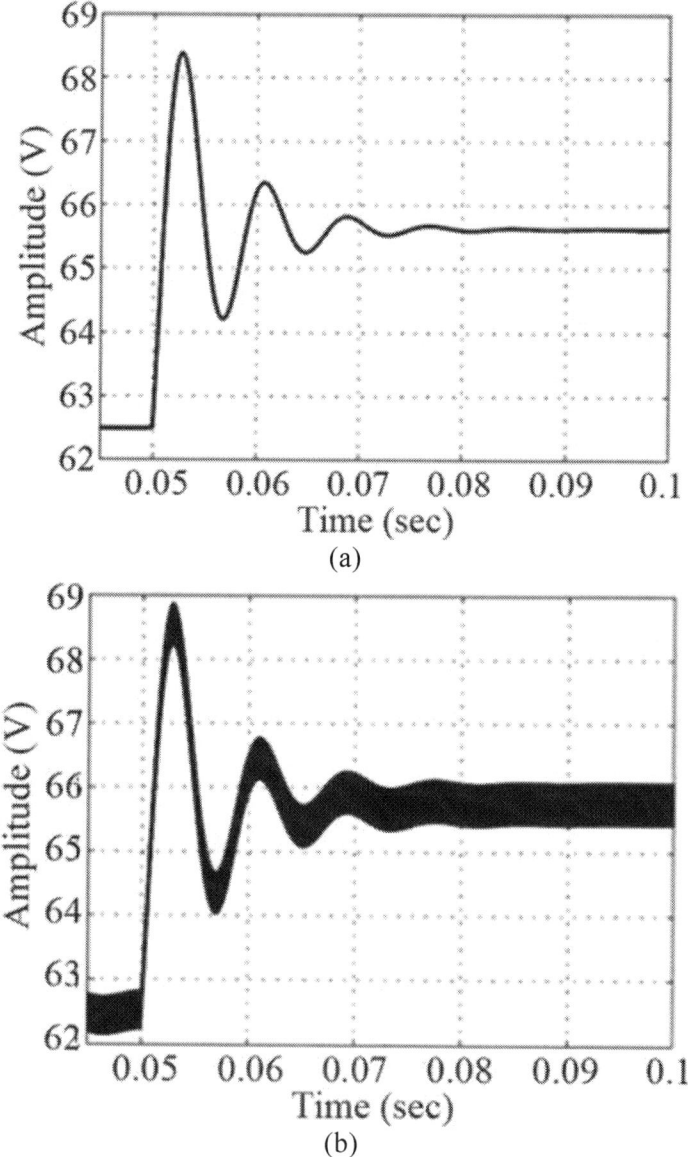

Figure 10. Inductor current in the presence of 0.01 step change in duty cycle. (a) Predicted response by mathematical averaged model; (b) Detailed circuit simulation in matlab/simulink.

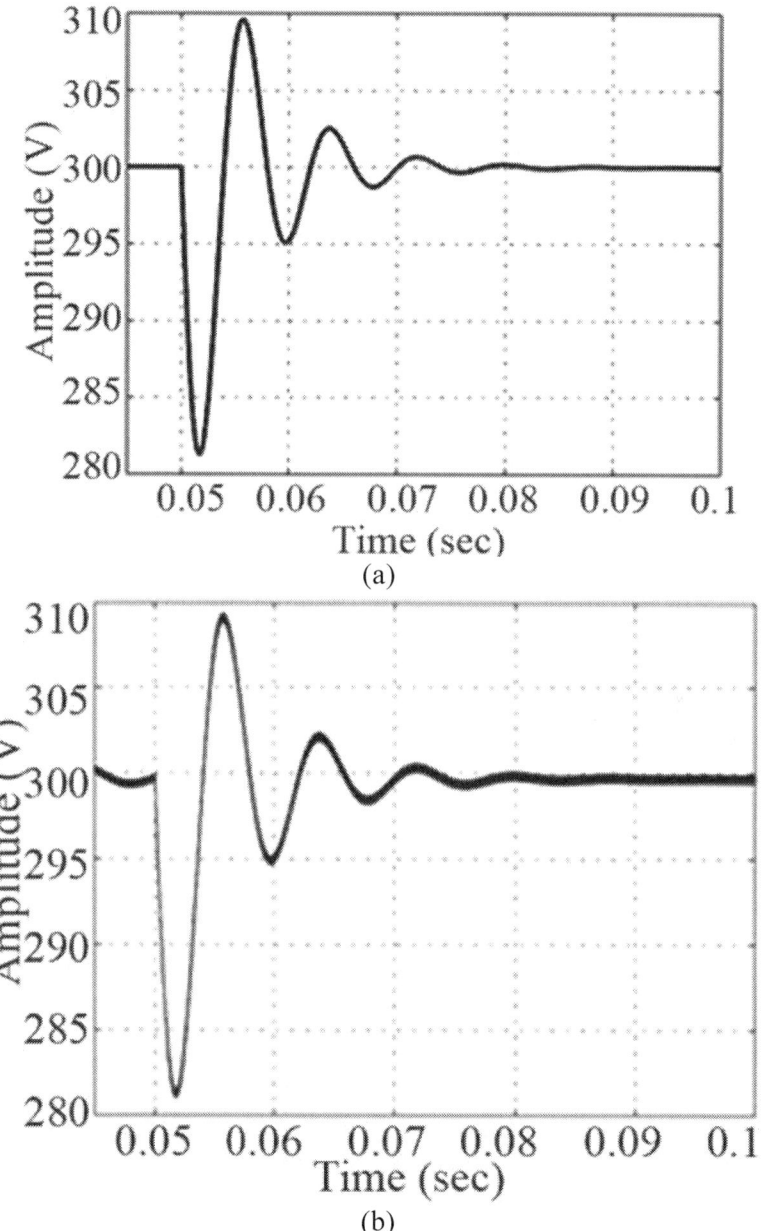

Figure 11. Output voltage in the presence of 1A step change in load current. (a) Predicted response by mathematical averaged model; (b) Detailed circuit simulation in matlab/ simulink.

Buck Mode

Start up process of open loop converter system with input voltage of 300V is simulated under these conditions: $D = 0.4$, $f_s = 20$ kHz, $L = 200$ μH, $C = 50$ μF, $V_{out} = V_c = 24$ V, $P_{in} = P_{out} = 1.5$ KW, $R = 0.384$ Ω.

From **Figure 12** it can be seen that the mathematical model has predicted the response of the converter very well. Figures 13 and 14 represent the behavior of output voltage and inductor current in the presence of 0.02 step change in duty cycle. **Figure 15** shows converter output voltage due to 0.5A step change in load, it is obvious that the mathematical model is in a close agreement with simulated circuit.

CONTROLLER DESIGN

Switch mode power supplies (SMPS) control methodology can be divided into two main methods consisting of voltage mode control (VMC) and current mode control. Current mode control (CMC) is faster than voltage mode but it suffers from ringing, so merging two methods can cover the shortage of both. Current programmed control (CP) method that is presented in [10,26,27] employs CMC and VMC together. **Figure 16** shows the block diagram of CPC which is mainly divided to peak, valley, and average [28]. Among all of those methods of CPC, the peak current programmed control (PCPC) is one of the most common modes and easiest one to understand.

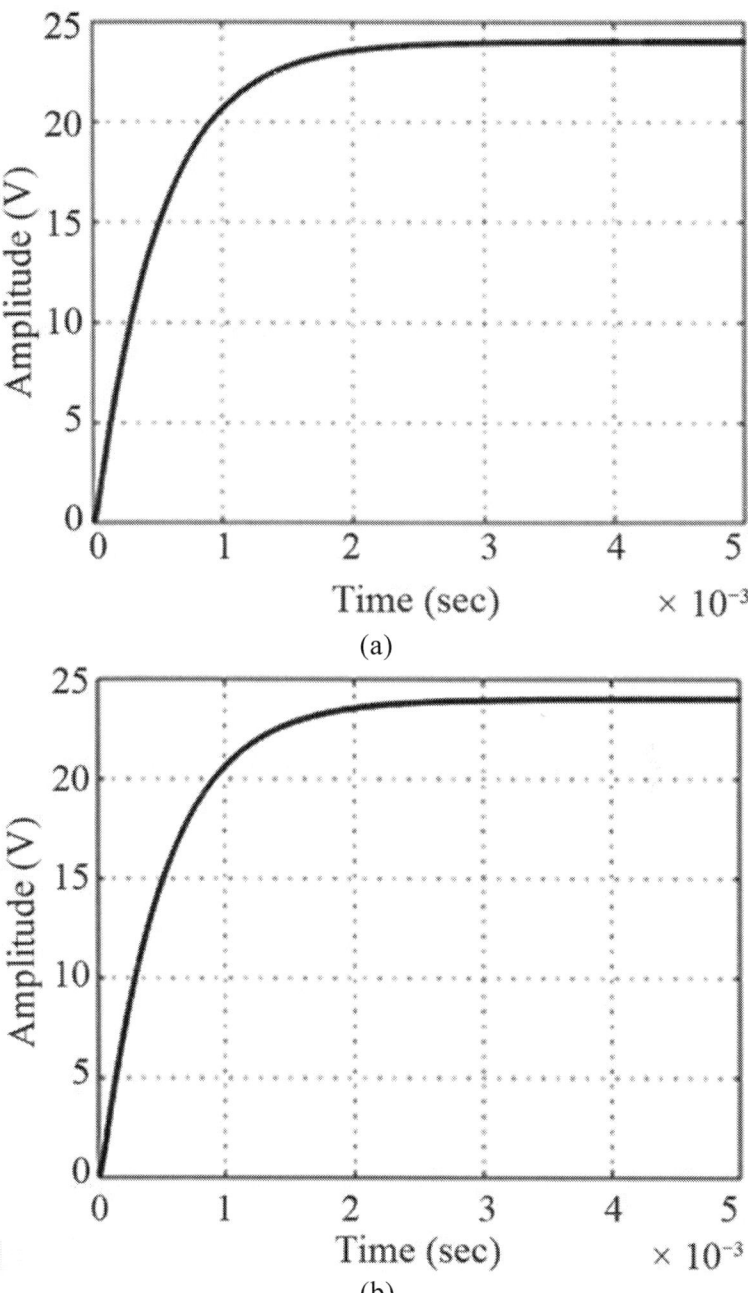

Figure 12. Open loop start up output voltage with fixed duty cycle (a) predicted by mathematical model; (b) Detailed circuit simulation with matlab/simulink.

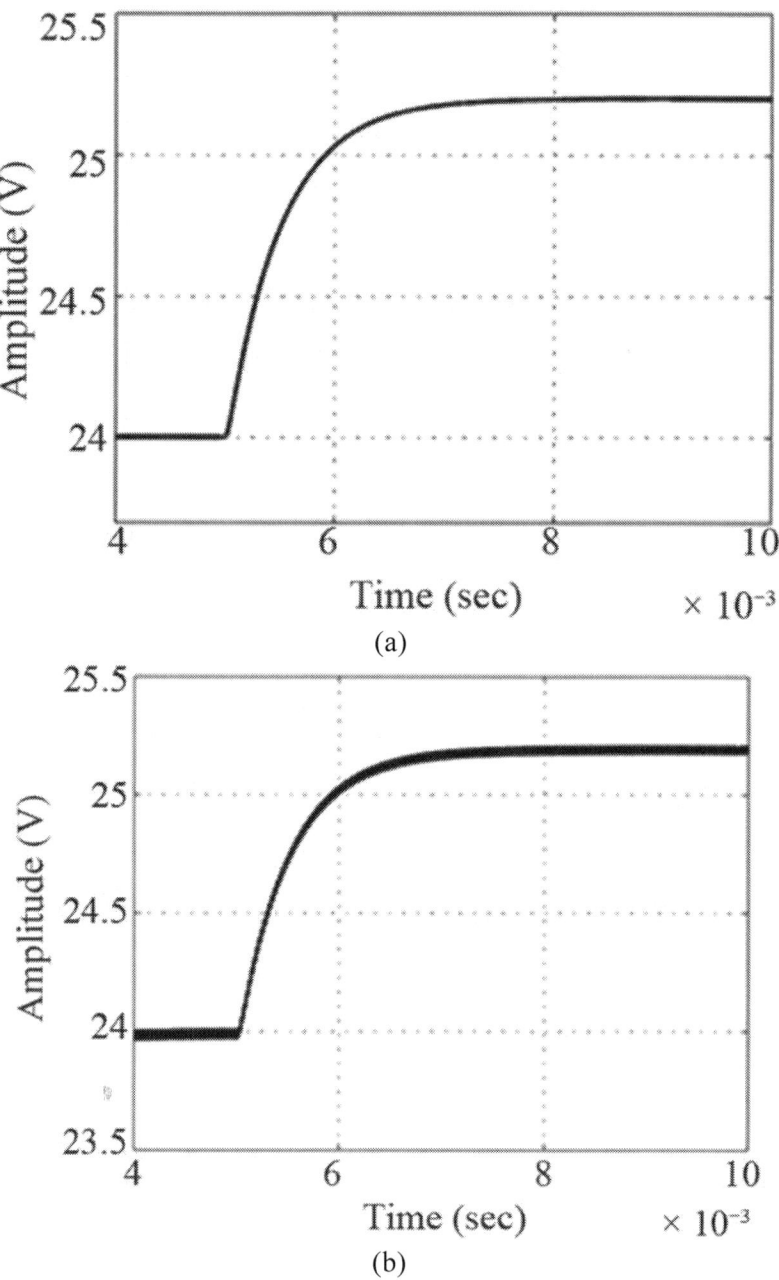

Figure 13. Open loop response of output voltage due to 0.02 step change in duty cycle (a) predicted by mathematical model; (b) Detailed circuit simulation with matlab/simulink.

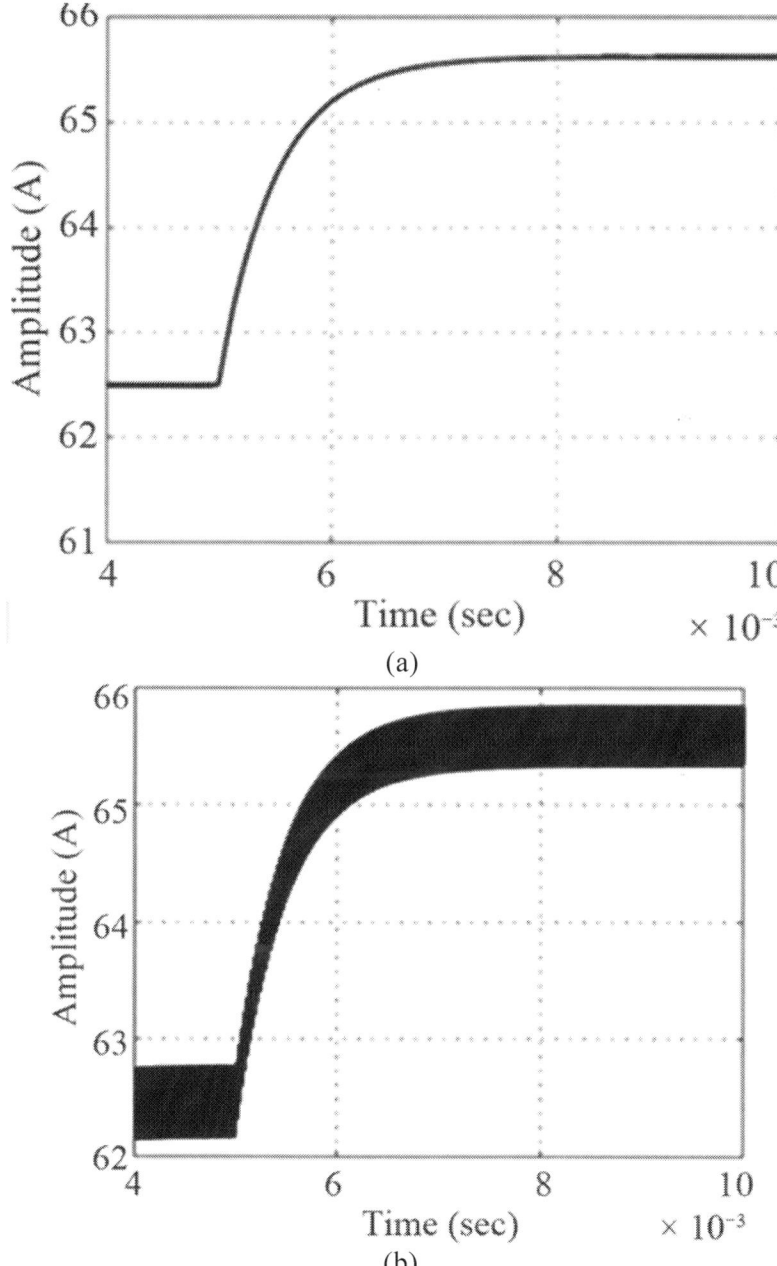

Figure 14. Open loop response of inductor current due to 0.01 step change in duty cycle (a) predicted by mathematical model; (b) Detailed circuit simulation with matlab/simulink.

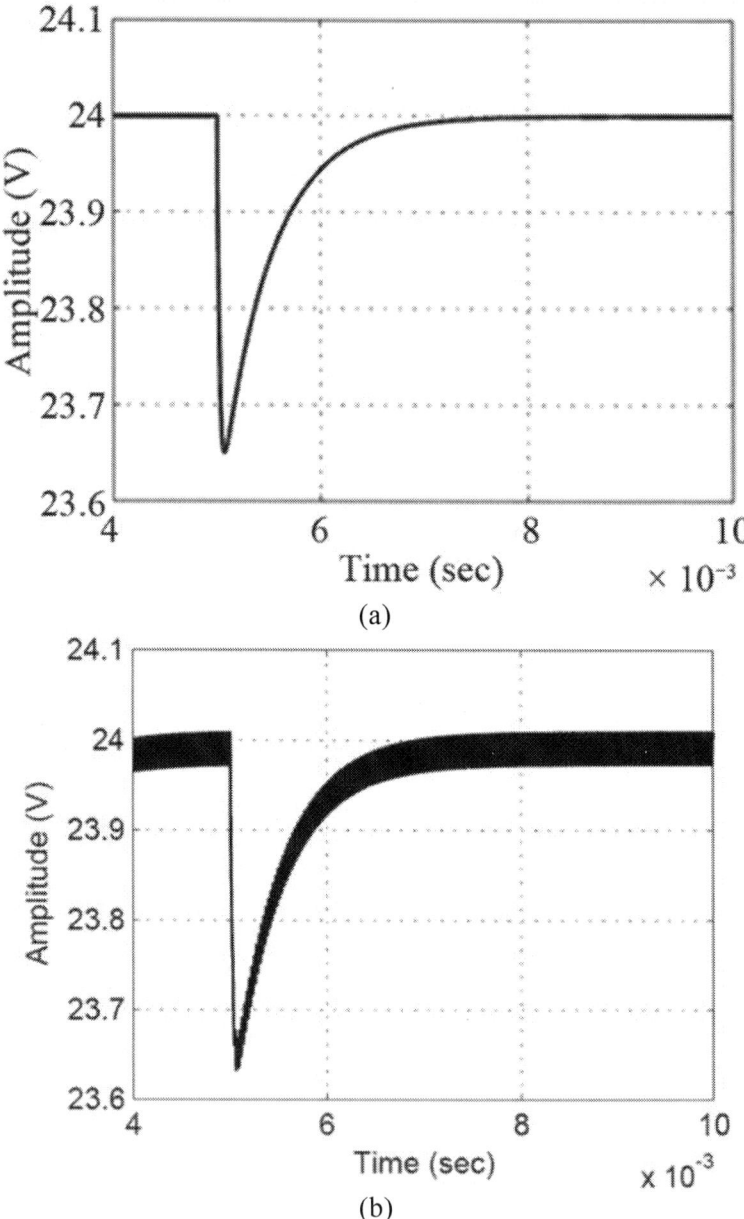

Figure 15. Open loop response of output voltage due to 1A step change load current (a) predicted by mathematical model; (b) Detailed circuit simulation with matlab/simulink.

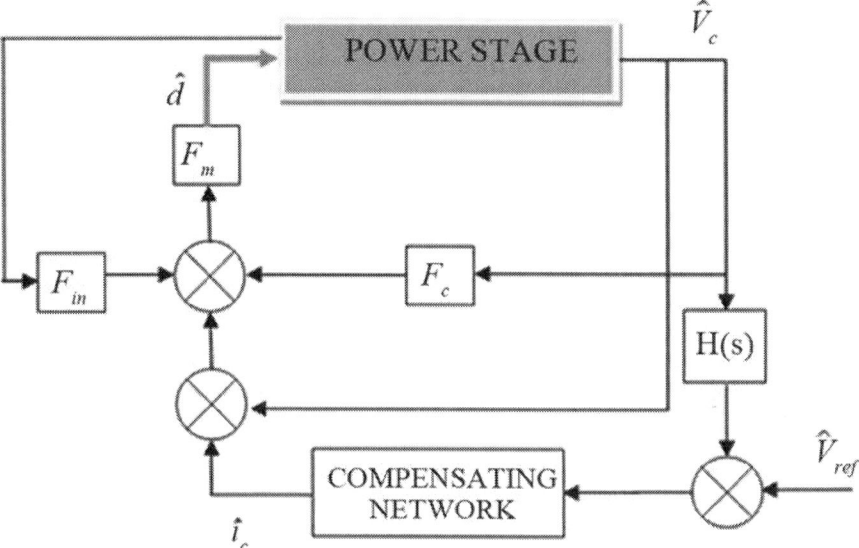

Figure 16. Block diagram of peak current mode control.

The method of PCMC modeling is the same as described in [10], only up and down slopes and their times have been replaced to adapt the formulation with proposed topology. So the model of **Figure 16** can be achieved, where F_m, F_{in} and F_c in boost and buck modes are determined in (26).

$$\tilde{d} = F_m \left[\tilde{i}_c - \tilde{i}_L - F_{in}\tilde{v}_{in} - F_c\tilde{v}_c \right]$$

$$\text{boost mode} \begin{cases} F_m = \dfrac{1}{M_a T_s} \\[2ex] F_{in} = \dfrac{(D-3/4)T_s}{L} \\[2ex] F_c = \dfrac{(1-D)^2 T_s}{nL} \end{cases}$$

(26)

$$\text{buck mode}\begin{cases} F_m = \dfrac{1}{M_a T_s} \\[3mm] F_{in} = \dfrac{D^2 T_s}{nL} \\[3mm] F_c = \dfrac{(1/4-D)T_s}{L} \end{cases}$$

Not considering input voltage and load variations, the overall block diagram for both modes of operations can be depicted as **Figure 17** with some algebraic operations; the loop gain transfer function of (27) can be concluded.

$$\text{loop gain} = H(S)G_c(S)\frac{F_m G_{V_c d}}{1+\beta F_m G_{id} + F_m F_c G_{V_c d}}$$

(27)

In order to examine the stability and design the converter's controller, boost and buck mode transfer functions can be replaced to achieve loop gain of each mode. With some algebraic operations one can get to (28) and (29) for boost and buck loop gain respectively. **Figure 18** shows the uncompensated and compensated loop gain of boost mode. Compensating network for boost mode operation is a simple PID which its coefficients have been selected with respect to the method presented by Chien and Fruehauf [29].

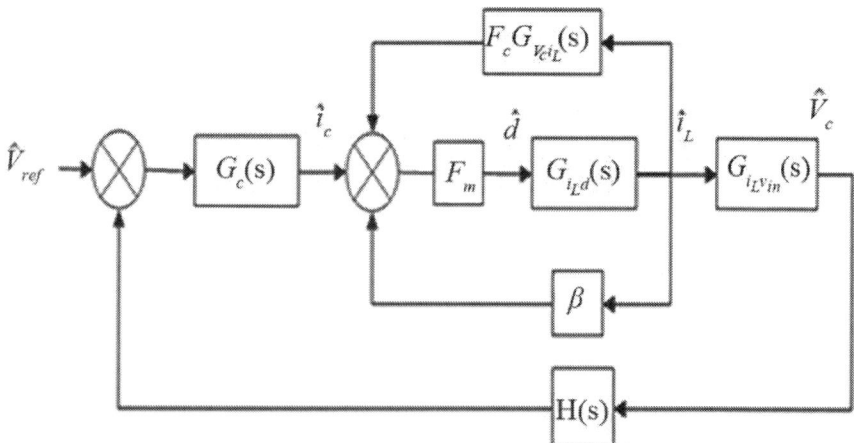

Figure 17. Simplified block diagram of closed loop converter with neglecting the effect of line and load variation.

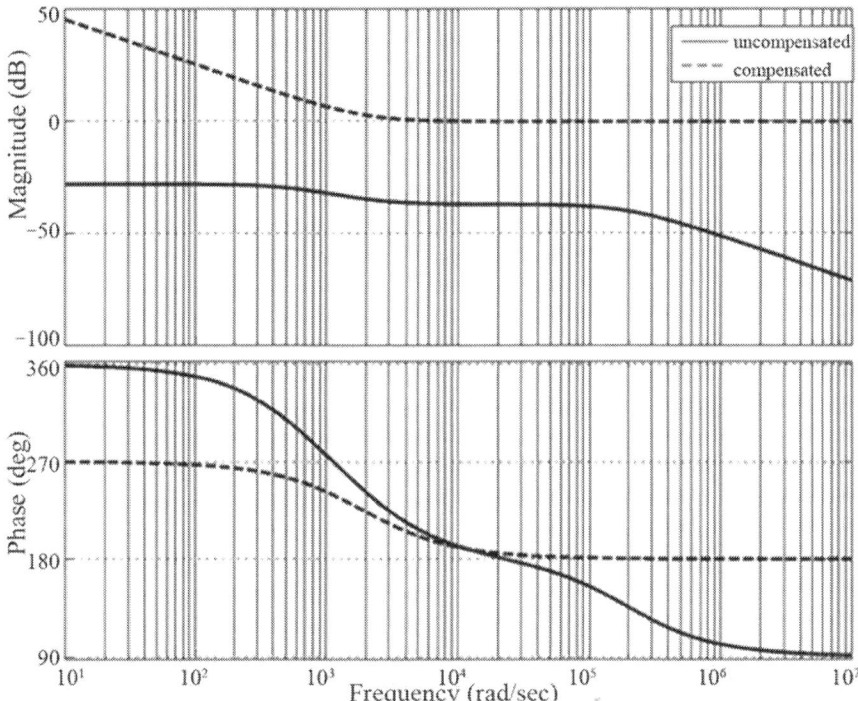

Figure 18. Uncompensated and compensated loop gain of boost mode solid line is uncompensated and dashed line is compensated response.

$$T_{boost}(S) = \frac{G_c(S)F_m H(S)(-2nLI_L S + 8(1-D)V_C - 2nV_{IN})}{den(S)}$$

$$den(S) = n^2 CLS^2 + \left(n^2\frac{L}{R} + 2nC\beta F_{mV_C} - 2nLI_L F_m F_c\right)S$$

$$+ 4(1-D)^2 + 8(1-D)F_m[\beta I_L + F_c V_C] - 2nF_m F_c V_{IN} \quad (28)$$

$$T_{buck}(S) = \frac{G_c(S)F_m H(S)[2nV_{IN}]}{den(S)}$$

$$den(S) = n^2 CLS^2 + \left(n^2\frac{L}{R} + 2nC\beta F_m V_{IN}\right)S$$

$$+ 2nF_m[\beta/R + F_c]V_{IN} \quad (29)$$

A simple integral controller can meet the requirements of buck mode compensating network. However, the designed controllers are not the optimum ones in terms of robustness. Other advanced control algorithms can be applied to this converter easily with the derived small signal model but that is beyond the scope of this work. **Figure 19** represents uncompensated and compensated loop gain of buck mode.

SIMULATION RESULTS

In order to make sure that the designed controllers are capable of controlling the converter, they should be examined. In control loop additional blocks like duty cycle limiter are employed because at the start up there is no output voltage. This leads to 100% the duty cycle, as a result after some cycles; the circuit is gone to its steady state in which inductor and capacitor experience only a constant source and this is when they go under short and open circuit respectively. The other important block is compensating ramp which can overcome the unstable oscillation problem described in [10,22,30]. Figures 20 and 21 show that the controllers are successful to control the converter in the presence of 1 A and 10 A step load change for boost and buck mode respectively.

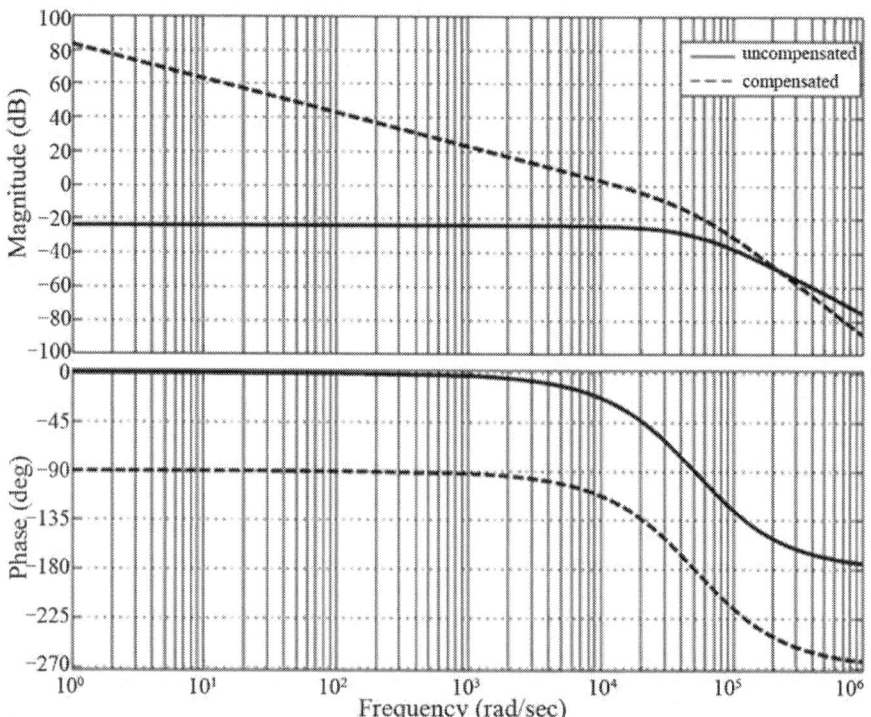

Figure 19. Uncompensated and compensated loop gain of buck mode, slid line is uncompensated and dashed line is compensated response.

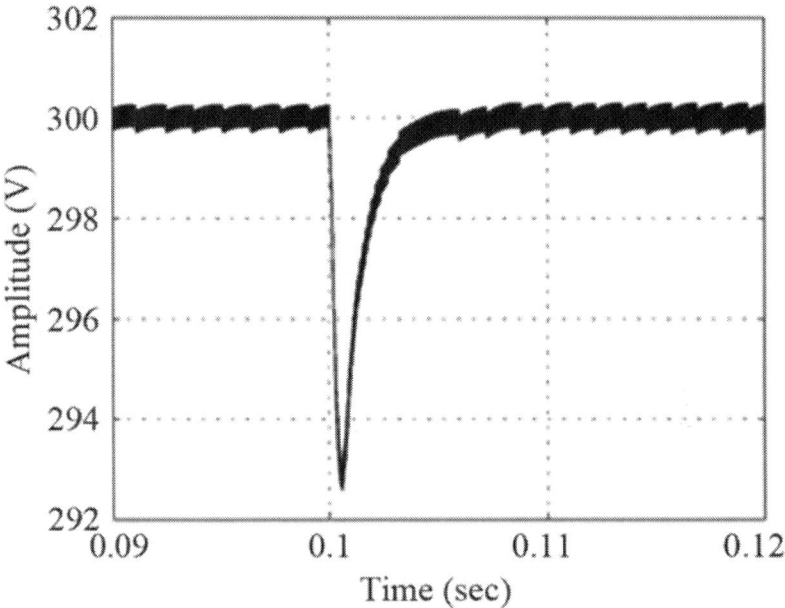

Figure 20. Response of compensated closed loop circuit in boost mode operation in the presence of 1 A step load change.

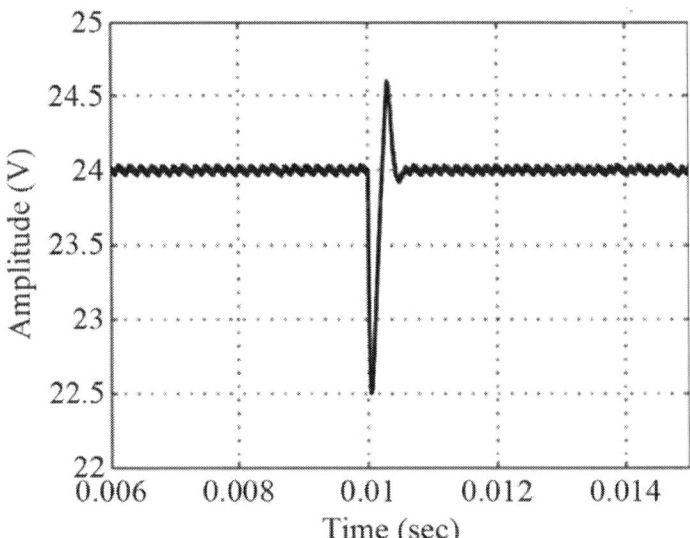

Figure 21. Response of compensated closed loop circuit in buck mode operation in the presence of 5 A step load change.

CONCLUSION

In this paper a bidirectional full bridge dc-dc converter as one of the most applicable industrial converters has been modeled with average state space method. Simulation results are in a close agreement with mathematical model which approve the successfulness of average state space to predict the response. Current programmed controller was designed for each mode. Although compensated network was effective in buck mode but employing modern and robust control algorithms can cope with the zero half plane problems in boost mode to achieve all the requirements such as more bandwidth, phase margin and capability of line and load change rejection.

REFERENCES

1. T. Mishima, E. Hiraki, T. Tanaka and M. Nakaoka, "A New Soft-Switched Bidirectional DC-DC Converter Topology for Automotive High Voltage DC Bus Architectures," IEEE Vehicle Power and Propulsion Conference, Windsor, 6-8 September, 2006, pp. 1-6.doi:10.1109/VPPC.2006.364281

2. A. E. Navarro, P. Perol, E. J. Dede and F. J. Hurtado, "A New Efficiency Low Mass Bidirectional Battery Discharger-Charger Regulator for Low Voltage Batteries", 27th Annual IEEE Power Electronics Specialists Conference, Baveno, 23-27 June 1996, pp. 842-845. doi:10.1109/PESC.1996.548679

3. E.-S. Kim, K.-Y. Joe, H.-Y. Choi, Y.-H. Kim and Y.-H. Cho, "An Improved Soft Switching Bi-Directional PSPWM FB DC/DC converter," Proceedings of the 24th Annual Conference of Industrial Electronics Society, Aachen, 31 August-4 September1998, pp. 740-743.doi:10.1109/IECON.1998.724185

4. H. L. Chan, K. W. E. Cheng and D. Suatanto, "A Novel Square-Wave Converter with Bidirectional Power Flow," Proceedings of the International Conference on Power Electronics and Drive Systems, Hohg kong, 27-29 July 1999, pp. 966-971.doi:10.1109/PEDS.1999.792839

5. M. Jain, P. K. Jain and M. Daniele, "Analysis of A Bidirectional DC-DC Converter Topology for Low Power Application," IEEE Canadian Conference on Electrical and Computer Engineering, St. Johns, 25-28 May 1997, pp. 548-551. doi:10.1109/CCECE.1997.608283

6. K. Venkatesan, "Current Mode Controlled Bideirectional Flyback Converter," 20th Annual IEEE Power Electronics Specialists Conference, Milwaukee, 26-29 June 1989, pp. 835-842.doi:10.1109/PESC.1989.48567

7. T. Reimann, S. Szeponik, G. Berger and J. Petzoldt, "A Novel Control Principle of Bidirectional DC-DC Power Conversion," 28th Annual IEEE Power Electronics Specialists Conference, St. Louis, 22-27 June 1997, pp. 978- 984. doi:10.1109/PESC.1997.616843

8. M. K. Kazimierczuk, D. Q. Vuong, B. T. Nguyen and J. A. Weimer, "Topologies of Bidirectional PWM DC-DC Power Converter," Proceedings of the IEEE National Aerospace and Electronics Conference, Dayton, 24-28 May 1993, pp. 435-441.doi:10.1109/NAECON.1993.290953

9. N. M. L. Tan, T. Abe and H. Akagi, "Topology and Application of Bidirectional isolated DC-DC Converters," IEEE 8th International Conference on Power Electronics and ECCE Asia, Jeju, 30 May-3 June 2011, pp. 1039- 1046. doi:10.1109/ICPE.2011.5944690

10. R. W. Erickson and D. Maksimovic, "Fundamental of Power Electronics," 2nd Edition, Kluwer Academic Publishers, Alphen aan den Rijn, 2001.

11. T. Mishima, E. Hiraki, T. Tanaka and M. Nakaoka, "High Frequency Link Symmetrical Active Edge Resonant Snubbers-assisted ZCS-PWM DC-DC Converter," Electric Power Applications, vol. 1, no. 6, 2007, pp. 907-914. doi:10.1049/iet-epa:20060508

12. C. Zhao, S. D. Round and J. W. Kolar, "Full-order Averaging Modeling of Zero-voltage Switching Phase-shift Bidirectional DC-DC Converters," Power Electronics, vol. 3, no. 3, 2010, pp. 400-410. doi:10.1049/iet-pel.2008.0208

13. F. Krismer and J. W. Kolar, "Accurate Small-signal Model for the Digital Control of an Automotive Bidirectional Dual Active Bridge," IEEE Transactions on Power Electronics, Vol. 24, No. 12, 2009, pp. 2756-2768. doi:10.1109/TPEL.2009.2027904

14. P. Jain and J. E. Quaicoe, "Generalized Modeling of Constant Frequency DC/DC Resonant Converter Topologies," 14th International Telecommunications Energy Conference, Washington DC, 4-8 October 1992, pp. 180-185. doi:10.1109/INTLEC.1992.268444

15. L. A. Aguirre, P. F. Donoso-Garcia and R. Santos-Filho, "Use of a Priori Information in the Identification of Global Nonlinear Models—A Case Study Using a Buck Converter," IEEE Transactions on Circuits and Systems I: Fundamental Theory and Applications, vol. 47, no. 7, 2000, pp. 1081-1085. doi:10.1109/81.855463

16. K. T. Chau and C. C. Chan, "Nonlinear Identification of Power Electronics Systems," Proceedings of International Conference on Power Electronics and Drive Systems, Singapore, 21-24 February 1995, pp. 329-334. doi:10.1109/PEDS.1995.404900

17. F. Alonge, F. D'Ippolito, F. M. Raimondi and S. Tumminaro, "Nonlinear Modeling of DC/DC Converters Using the Hammerstein's Approach," IEEE Transactions on Power Electronics, Vol. 22, No. 4, 2007, pp. 1210-1221. doi:10.1109/TPEL.2007.900551

18. F. Alonge, F. D'Ippolito and T. Cangemi, "Hammerstein Model-Based Robust Control of DC/DC Converters," 7th International Conference on Power Electronics and Drive Systems, Bangkok, 27-30 November 2007, pp. 754-762. doi:10.1109/PEDS.2007.4487788

19. V. Vorperian, R. Tymerski and F. C. Lee, "Equivalent Circuit Model for Resonant and PWM Switches," IEEE Transactions on Power Electronics, vol. 4, no. 2, 1989, pp. 205-214.doi:10.1109/63.24905

20. V. Vorperian, "Simplified Analysis of PWM Converter Using the Model of the PWM Switch: part I and II," IEEE Transactions on Aerospace and Electronic Systems, vol. 26, no. 3, 1990, pp. 497-505. doi:10.1109/7.106127

21. R. D. Middlebrook and S. Cuk, "A General Unified Approach to Modeling Switching Converter Power Stages," International Journal of Electronics, vol. 42, no. 6, 1977, pp. 521-550. doi:10.1080/00207217708900678

22. S.-P. Hsu, A. Brown, L. Rensink and R. D. Middle-brook, "Modeling and Analysis of Switching DC-to-DC Converters in Constant-Frequency Current Programmed Mode", Power Electronics specialists Conference, San Diego, 18-22 June 1979, pp. 284-301.

23. S. Cuk, "Modeling, Analysis, and Design of Switching Converters," Ph.D. thesis, California Institute of Technology, Pasadena, 1976.

24. L. Zhu, "A Novel Soft-Commutating Isolated Boost Full-Bridge ZVS-PWM DC-DC Converter for Bidirectional High Power Applications," IEEE Transactions on Power Electronics, Vol. 21, No. 2, 2004, pp. 422-429. doi:10.1109/TPEL.2005.869730

25. R. Li , A. Pottharst, N. Frohleke and J. Bocker, "Analysis and Design of Improved Isolated Full-Bridge Bidirectional DC-DC Converter," 35th Annual Power Electronics Specialists Conference, Aachen, 20-25 June 2004, pp. 521-526. doi:10.1109/PESC.2004.1355801

26. R. B. Ridley, "A New, Continuous-Time Model for Current Mode Control," IEEE Transactions on Power Electronics, vol. 6, No. 2, 1991, pp. 271-280.doi:10.1109/63.76813

27. L. Peng and B. Lehman, "A Design Method for Paralleling Current Mode Controlled DC-DC Converters," IEEE Transactions on Power Electronics, Vol. 19, no. 3, 2004, pp 748-756.doi:10.1109/TPEL.2004.826497

28. R. Sheehan, "Understanding and Applying Current-Mode Control Theory," Power Electronics Technology Exhibition and Conference, Dallas, 30 October-1 November,2007, pp. 1-26.

29. D. E. Seborg, T. F. Edgar and D. A. Millichamp, "process dynamics and control," 2nd Edition, John Wiley and Sons Inc., New York, 2004.
30. L. Dixon, "Average Current Mode Control of Switching Power Supplies," Unitrode application note, No. 3, pp. 356-369.

CITATION

S. Moghaddam, A. Ayatollahi and A. Rahmati, "Modeling and Current Programmed Control of a Bidirectional Full Bridge DC-DC Converter," Energy and Power Engineering, Vol. 4 No. 3, 2012, pp. 107-116. doi: 10.4236/epe.2012.43015.

CHAPTER 6

PI and RST Control Design and Comparison for Matrix Converters Using Venturini Modulation Strategy

Bekhada Hamane[1], Mamadou Lamine Doumbia[1], Hicham Chaoui[2], Mohamed Bouhamida[3], Ahmed Chériti[1] and Mustapha Benghanem[3]

[1]Department of Electrical and Computer Engineering, UQTR, Trois-Rivières, Canada
[2]Center for Energy Systems Research, Department of Electrical and Computer Engineering, Tennessee Technological University, Cookeville, USA
[3]Department of Electrical Engineering, University Mohamed Boudiaf, Oran, Algeria

ABSTRACT

This paper presents a thorough design and comparative study of two popular control techniques, i.e., classical Proportional Integral (PI) and RST, for Matrix Converters (MCs) in terms of tracking the reference and robustness. The output signal of MCs is directly affected by unbalanced grid voltage. Some research works have attempted to overcome this problem with PI control. However, this technique is known to offer lower performance when it is used in complex and nonlinear systems. On the other hand, RST control offers better performance, even in case of highly nonlinear systems. Therefore, the RST can achieve better performance to overcome the limitation of PI control of nonlinear systems. In this paper, a RST control method is proposed as output current controller to improve the performance of the MC powered by unbalanced grid voltage. The overall operating principle, Venturini modulation strategy of MC, PI control and characteristics of RST are presented.

INTRODUCTION

Recent advances in power electronics have enabled the emergence of Matrix Converter (MC) for direct AC/AC conversion [1] . Interest in this converter topology was rather academic with efforts provided in many research laboratories [1]. MC uses bidirectional current and voltage power switches that connect converter input and output phases [2] . The direct conversion is performed without intermediate DC link circuit for energy storage [2] [3] . MC was introduced firstly in 1976. To prevent the spread of current harmonics caused by the MC to the supply network, an input LC filter is used. It provides a very low impedance path and absorbs current harmonics [1] [2]. Venturini and Alesina proposed a generalized high-frequency switching strategy in 1980 [3] . The objective of this control strategy is to achieve an ideal electronic transformer capable of varying the voltage, current, frequency and power factor [4] . Another method, known as the direct transfer function approach, proposes the multiplication of the input voltages vectors by the modulation matrix M to obtain a vector of output voltages which correspond to a point of synthesis [4] . However, the simultaneous commutation of controlled bidirectional switches used in MC is very difficult to achieve without generating over current or overvoltage spikes which can destroy the power semiconductors [3] . Also, the load side of the MC is directly affected by the distorted and/or unbalanced input voltages due to the lack of DC intermediate circuit in the MC. The performance of the MC deteriorates, when it is exposed to the harmonic and non-sinusoidal currents and some papers have presented mitigation methods [3] [5] . Conventional PI controller works well only if the mathematical model of the system could be computed. However, it is difficult to implement the conventional PI controller for variable as well as complex systems [5] [6] . So, RST Controller is investigated. This regulator, whose synthesis is purely algebraic, is a sophisticated algorithm based on pole placement method which exploits many numerical resources [7] [8] . The method used to determine the gains of the PI controller is the compensation method of poles, we note here that the interest of the compensation of the poles occurs only if the system parameters are accurately identified as gains K_p and K_i are based on these same parameters. If the actual parameters are different from those used in the synthesis, the compensation is ineffective. In the literature, control law design approaches can be divided into two categories. The first category consists of a nonlinear systems linearization around an operating point of the states. In this case, classical linear control laws are applied for the approximated system. These methods are popular in the industry and are mainly used for their simplicity. However, the control system's performance and stability are not guaranteed for the overall system. The second category deals with nonlinear controllers design based on nonlinear systems dynamics. In this category, the characteristics of nonlinear systems are preserved. However, the design approach difficulties arise with the complexity of the nonlinear systems dynamics. Furthermore, these approaches assume a precise mathematical system model and are able to cope with nonlinearities to a certain degree. But, their performance also degrades in the presence of varying operating conditions, and higher uncertainties and disturbances. Therefore, this paper aims to compare the most popular techniques in the industry with similar design complexity. This work presents a modeling, theoretical analysis and an in-depth comparison of both the classical PI and RST Controller for MCs. Results show the superiority of the RST strategy with faster dynamic response and better robustness. To show the effectiveness of the control methods, the performance of the system is analyzed and compared in various operating conditions.

MATHEMATICAL MODEL OF MATRIX CONVERTER

This part consists of a brief description and modeling of each element of the matrix converter. We start with modeling the MC, then the input filter and it ends with the load RL. Ideal bidirectional switches are represented by S_{ij}, where $i=\{A,B,C\}$ and $j=\{a,b,c\}$ represent respectively the index of input and output voltage [1] [9] [11] :

$$S_{ij} = \begin{cases} 1 & \text{If the switch } S_{ij} \text{ is closed} \\ 0 & \text{If the switch } S_{ij} \text{ is opened} \end{cases}$$

(1)

$$S_{Aj} + S_{Bj} + S_{Cj} = 1$$

(2)

The basic diagram of a MC is represented in Figure 1, which the clipping circuit is used to protect the converter against surges that could come from a sudden disconnection of the load [1] .

With these restrictions, a 3×3 matrix converter has 27 possible switching states [1] . Let m_{ij} be the duty cycle of switch S_{ij}, defined as [1] [10] [11] :

$$m_{ij}(t) = \frac{t_{ij}}{T_{seq}}$$

(3)

where, $0 < m_{ij} < 1$, $T_{seq} = \dfrac{1}{f_s}$ and f_s is the switching frequency.

The transfer matrix of the converter is defined by [1] [3] [10] [11] :

$$M = \begin{bmatrix} m_{Aa} & m_{Ba} & m_{Ca} \\ m_{Ab} & m_{Bb} & m_{Cb} \\ m_{Ac} & m_{Bc} & m_{Cc} \end{bmatrix} \tag{4}$$

Figure 2 shows an example of the duration of conduction of the switches during a switching sequence T_{seq} of the MC [1] [12] .

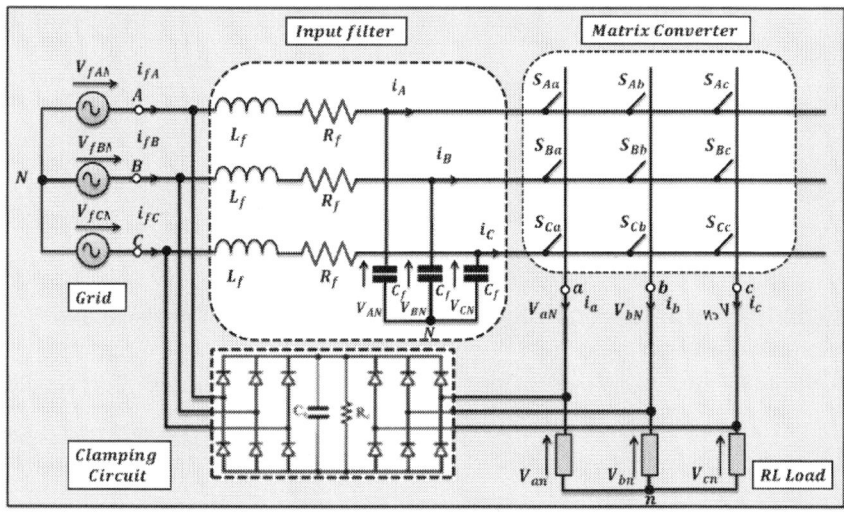

Figure 1. Basic circuit of a Matrix Converter.

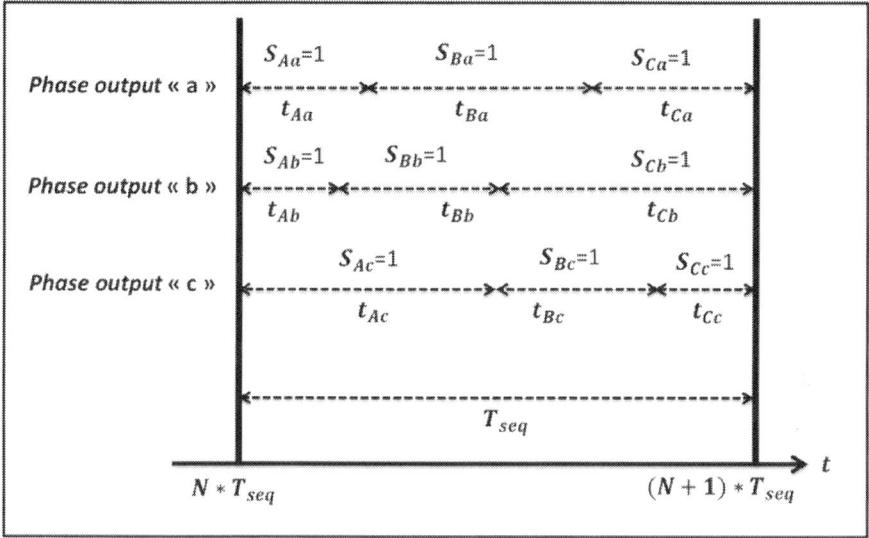

Figure 2. Example of the operation timing of switches during a switching period.

Modeling of the Matrix Converter

The input voltage and current of the matrix converter are given by [1] [10] [11] [13] :

$$V_i = V_{im} \begin{bmatrix} \cos(\omega_i t) \\ \cos\left(\omega_i t + \dfrac{2\pi}{3}\right) \\ \cos\left(\omega_i t + \dfrac{4\pi}{3}\right) \end{bmatrix}$$

(5)

$$I_i = I_{im} \begin{bmatrix} \cos(\omega_i t + \varphi_i) \\ \cos\left(\omega_i t + \varphi_i + \dfrac{2\pi}{3}\right) \\ \cos\left(\omega_i t + \varphi_i + \dfrac{4\pi}{3}\right) \end{bmatrix}$$

(6)

Assuming the relationship between the output and the input signal of the matrix converter [1] [10] [11] [14] :

$$q = \sqrt{\frac{V_o^2}{V_i^2}} = \sqrt{\frac{I_i^2}{I_o^2}} \quad \text{with}: 0 < q \le 0.866 \tag{7}$$

The matrix converter will be designed and controlled to provide desired output voltage and output current [1] [10] [11] [13] :

$$V_o = V_{om} \begin{bmatrix} \cos(\omega_o t) \\ \cos\left(\omega_o t + \dfrac{2\pi}{3}\right) \\ \cos\left(\omega_o t + \dfrac{4\pi}{3}\right) \end{bmatrix} \tag{8}$$

$$I_o = I_{om} \begin{bmatrix} \cos(\omega_o t + \varphi_o) \\ \cos\left(\omega_o t + \varphi_o + \dfrac{2\pi}{3}\right) \\ \cos\left(\omega_o t + \varphi_o + \dfrac{4\pi}{3}\right) \end{bmatrix} \tag{9}$$

The neutral to phase output voltages V_{an}, V_{bn} and V_{cn} are given by [1] [3] [9] [11] :

$$\begin{bmatrix} V_{an} \\ V_{bn} \\ V_{cn} \end{bmatrix} = \begin{bmatrix} m_{Aa} & m_{Ba} & m_{Ca} \\ m_{Ab} & m_{Bb} & m_{Cb} \\ m_{Ac} & m_{Bc} & m_{Cc} \end{bmatrix} \begin{bmatrix} V_{AN} \\ V_{BN} \\ V_{CN} \end{bmatrix} \tag{10}$$

The input current I_A, I_B and I_C are [3] [9] [11] :

$$\begin{bmatrix} I_A \\ I_B \\ I_C \end{bmatrix} = \begin{bmatrix} m_{Aa} & m_{Ab} & m_{Ac} \\ m_{Ba} & m_{Bb} & m_{Bc} \\ m_{Ca} & m_{Cb} & m_{Cc} \end{bmatrix} \begin{bmatrix} I_a \\ I_b \\ I_c \end{bmatrix} \tag{11}$$

ω_i, V_{im} are respectively the input voltage frequency and amplitude;

I_{im}, φ_i are respectively the input current amplitude and input phase;

ω_o, V_{om} are respectively the output voltage frequency and amplitude.

Modeling of the Input Filter

The LC input filter [15] (represented as shown in Figure 3) is a series resonant circuit tuned to the frequency of harmonics and connected in shunt. It provides a very low impedance path and absorbs harmonic currents [1] [3] [14] . At the fundamental frequency, the filter acts as a reactive power compensator [1] [3] . The LC input filter may be modeled with the equivalent circuit [15] . From the Kirchhoff's laws, node equations and Laplace transformation.

The filter output voltage and input current are obtained as Equation (12) and Equation (13) [1] [10] [12] .

$$V_{AN}(p) = \frac{1}{L_f C_f p^2 + R_f C_f p + 1} V_{fAN} - \frac{L_f p + R_f}{L_f C_f p^2 + R_f C_f p + 1} I_A \tag{12}$$

$$I_{fA}(p) = \frac{1}{L_f C_f p^2 + R_f C_f p + 1} I_A + \frac{C_f p}{L_f C_f p^2 + R_f C_f p + 1} V_{fAN} \tag{13}$$

Modeling of the Load RL

Generally, the neutral at the load (n) is isolated from that of the source (N) as shown in Figure 1. Therefore, the objective is calculating the load current, it is necessary to know the potential at the output of the MC corresponding to the neutral of the load. In this case, we have [1] [16] :

$$V_{jn} = V_{jN} - V_{nN} \tag{14}$$

The potential difference between the two neutral is given by [1] [16] :

$$V_{nN} = \frac{V_{aN} + V_{bN} + V_{cN}}{3} \tag{15}$$

As the transfer function of the load current is given by [1] [16] :

$$i_j(p) = \frac{1}{L_i p + R_i} V_{jn}(p)$$

(16)

VENTURINI MODULATION STRATEGY OF MATRIX CONVERTER

This method can produce the sinusoidal input current with unity power factor independently of load [4] [9] . The principle is to synthesize the desired three-phase output voltage from the input during each defined switching period. The initial equations of Venturini method are obtained as the product the ratio q, the voltage amplitude, third harmonic frequency of the input and output voltage as indicated in references [3] [10] [17] :

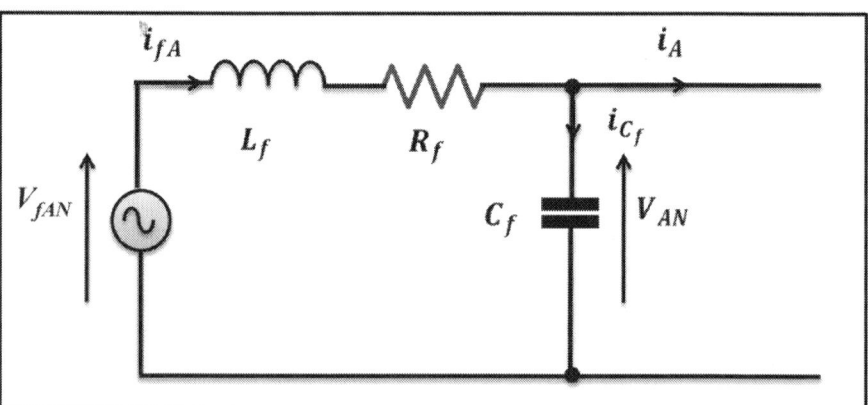

Figure 3. Input filter scheme.

$$V_o(t) = q V_{im} \begin{bmatrix} \cos(\omega_o t) - \frac{1}{6}\cos(3\omega_o t) + \frac{1}{2\sqrt{3}}\cos(3\omega_i t) \\ \cos\left(\omega_o t + \frac{2\pi}{3}\right) - \frac{1}{6}\cos\cos(3\omega_o t) + \frac{1}{2\sqrt{3}}\cos\cos(3\omega_i t) \\ \cos\left(\omega_o t + \frac{4\pi}{3}\right) - \frac{1}{6}\cos\cos(3\omega_o t) + \frac{1}{2\sqrt{3}}\cos\cos(3\omega_i t) \end{bmatrix}$$

(17)

According to the optimal amplitude in expression of Venturini, the modulation function is [1] [11] [14] [17] :

$$m_{ij} = \frac{1}{3}\left[1 + \frac{2V_i V_j}{V_{im}^2}\right]$$

(18)

The S_{ij} can be obtained according to the logic rules using the activation times t_{ij} [11] [17] , as shown inFigure 4.

Therefore, only six duty cycles are sufficient to calculate the gate signals of the power switches [10] [11] [13] .

$$\left.\begin{aligned} X &= t_{Aj} \\ Y &= t_{Aj} + t_{Bj} \end{aligned}\right\} \Rightarrow \begin{cases} S_{Aj} = (X) \\ S_{Bj} = (\overline{X}) \text{ and } (Y) \\ S_{Cj} = (\overline{X}) \text{ and } (\overline{Y}) \end{cases}$$

(19)

The carrier signal is expressed by [10] [11] [13] :

$$U_p = \frac{1}{T_{seq}} t \quad \text{with} : 0 \le t \le T_{seq}$$

(20)

CONTROL DESIGN

This section deals with the design and synthesis of the PI and RST controllers. Both controllers are designed to achieve current reference tracking with constant and varying current reference signals. This also has to be achieved under both balanced and unbalanced grid voltage conditions.

PI Controller Design

Current measurements of the load RL using a PI controller is illustrated by Figure 5.

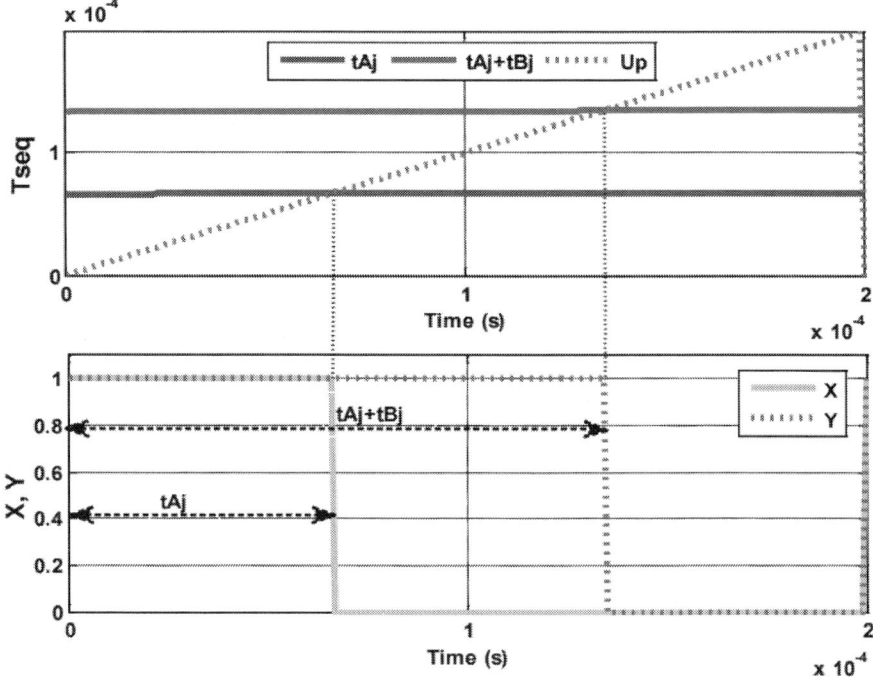

Figure 4. Obtaining logical instructions X and Y.

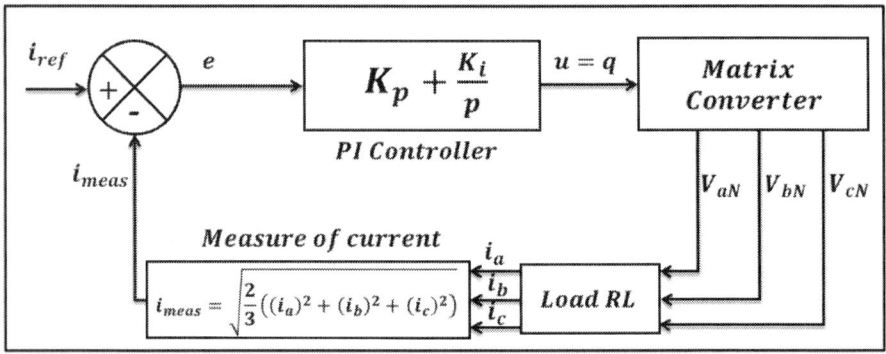

Figure 5. PI Controller for matrix converter.

The transfer function of the system is:

$$T(p) = \frac{B}{A} = \frac{1}{R_l + L_l p}$$

(21)

The values of A and B are:

$$\begin{cases} A = R_l + L_l p \\ B = 1 \end{cases}$$

(22)

The transfer function of the open-loop including the regulator is:

$$G(p) = \left(\frac{p + \dfrac{K_i}{K_p}}{\dfrac{p}{K_p}} \right) \cdot \left(\frac{\dfrac{1}{L_l}}{p + \dfrac{R_l}{L_l}} \right)$$

(23)

To cancel the pole, a zero was added at the same location as the pole [18].
Equation (24) gives a pole value:

$$\frac{K_i}{K_p} = \frac{R_l}{L_l}$$

(24)

The transfer function of the open-loop becomes:

$$G(p) = \frac{K_p \dfrac{1}{L_l}}{p}$$

(25)

The transfer function of the closed loop is expressed by:

$$H(p) = \frac{1}{1 + p\tau_r}$$

(26)

Which:

$$\tau_r = \frac{L_l}{K_p}$$

(27)

For a response time τ_r=0.66s, the K_p and K_i can be expressed by,

$$\begin{cases} K_p = \dfrac{L_l}{\tau_r} \\ K_i = \dfrac{R_l}{\tau_r} \end{cases}$$

(28)

RST Controller Design

The closed-loop system of the RST controller for MC is given by the following block diagram in Figure 6.

The goal of this section to determinate the RST controller's current. This type of controller is a structure with two freedom degrees and compared to a one degree of freedom structure, it has the main advantage that it allows the designer to specify performances independently with reference trajectory tracking (reference variation) and with regulation [7] [17] . It is based on the pole placement theory [8] , which consists in specifying an arbitrary stability polynomial D(p) and calculate S(p)and R(p) according to the Bezout equation [7] [17] :

$$D = AS + BR$$

(29)

With:

$$\deg(D) = \deg(A) + \deg(S)$$

(30)

For our model, we obtain [17] :

$$\begin{cases} A = a_1 p + a_0 \\ B = b_0 \\ D = d_3 p^3 + d_2 p^2 + d_1 p + d_0 \\ R = r_1 p + r_0 \\ S = s_2 p^2 + s_1 p + d_0 \end{cases}$$

(31)

The terms A and B are expressed by Equation (22). According to the robust pole placement strategy [8] , the polynomial D is written as [17] :

$$D = \left(p + \frac{1}{T_c} \right)\left(p + \frac{1}{T_f} \right)^2$$

(32)

To accelerate the system, the following conditions were adopted:

$$D = (s - 5P_a)(s - 15P_a)^2$$

(33)

With $P_C = -1/T_C$ pole of polynomial order and $P_f = -1/T_f$ double pole of the polynomial filter F [17] .

$$\begin{cases} P_c = 5P_a = -5\dfrac{R_l}{L_l} \\ T_c = \dfrac{1}{P_c} \\ T_f = \dfrac{1}{3}T_c \end{cases}$$

(34)

By identifying Equation (31) and Equation (34), coefficients of polynomial D were found and are linked to the coefficients of R and S by the Sylvester Matrix [7] [17] . Thus, the parameters of the RST controller can be determined as follows:

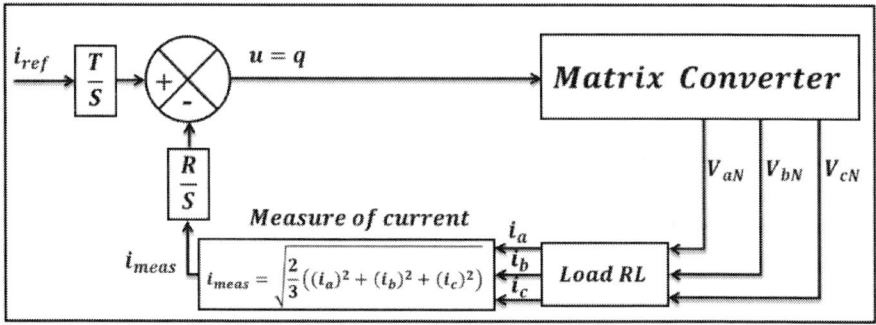

Figure 6. RST Controller for matrix converter.

$$
\begin{cases}
d_3 = a_1 s_2 \rightarrow s_2 = \dfrac{d_3}{a_1} \\[2mm]
d_2 = a_1 s_1 \rightarrow s_1 = \dfrac{d_2}{a_1} \\[2mm]
d_1 = a_0 s_1 + b_0 r_1 \rightarrow r_1 = \dfrac{d_1 - a_0 s_1}{b_0} \\[2mm]
d_1 = b_0 r_0 \rightarrow r_0 = \dfrac{d_0}{b_0} \\[2mm]
T = r_0
\end{cases}
$$

$$(35)$$

The reference current is calculated as shown in Figure 7 [13] [18] .
The measured load's current and the reference load's current are given by
Equation (36) [13] [18] :

$$
\begin{cases}
i_{meas} = \sqrt{\dfrac{2}{3}\left(\left(i_a\right)^2 + \left(i_b\right)^2 + \left(i_c\right)^2\right)} \\[4mm]
i_{ref} = \sqrt{\dfrac{2}{3}\left(\left(i_{aref}\right)^2 + \left(i_{bref}\right)^2 + \left(i_{cref}\right)^2\right)}
\end{cases}
$$

$$(36)$$

SIMULATIONS RESULTS

The PI and RST are used to control a matrix converter and a set of simulation runs is performed using SimPowerSystems toolbox of Matlab/Simulink software. The input filter parameters are calculated as given in [14] . Bidirectional switches MOSFET are considered ideal and ode23tb simulation solver was used. The MC system's parameters are listed in Table 1.

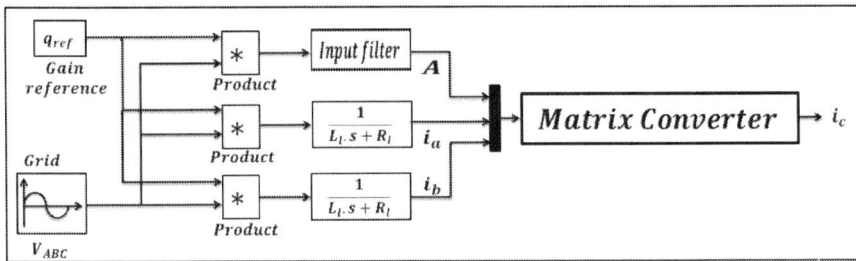

Figure 7. Load reference current.

Table 1. System Parameters.

Parameters	Values
Input voltage phase to neuter RMS	$V_{im} = 220$ V
Input frequency	$f_i = 50$ Hz
Switching frequency	$f_s = 5$ KHz
Input filter resistance	$R_f = 0.08\,\Omega$
Input filter inductance	$L_f = 30$ mH
Input filter capacitor	$C_f = 25\,\mu$F
Load resistance	$R_1 = 10\,\Omega$
Load inductance	$L_1 = 55$ mH
Input voltage phase to neuter RMS	$V_{im} = 220$ V
Input frequency	$f_i = 50$ Hz

Balanced Grid Case with PI Controller
Figure 8 shows the balanced grid voltage.

- Constant reference current I_{ref}:

Figure 9 shows the output voltage and linear load current using PI controller for balanced grid voltage with constant current reference. Figure 10 presents load current and variation of the ratio q. PI controller is used and the grid voltage balanced. Figure 11 shows the THD of load current with constant current reference.

- Time-varying reference current I_{ref}::

Figure 12 shows the output voltage and linear load current using PI controller for balanced grid voltage with stepped changing reference current. Figure 13 presents load current and variation of the ratio q. PI controller is used and the grid voltage balanced. Figure 14 shows the THD of load current with stepped changing reference current.

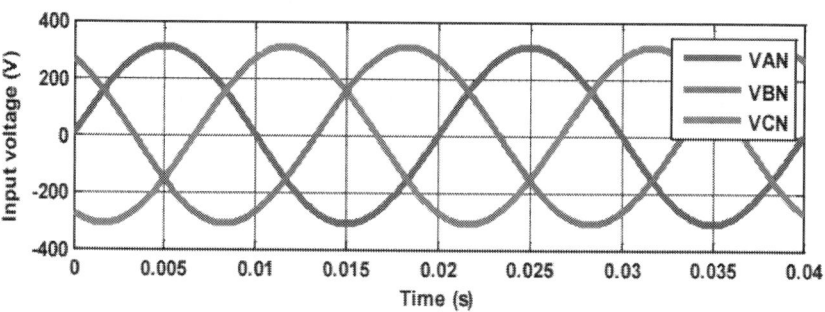

Figure 8. Balanced grid voltage.

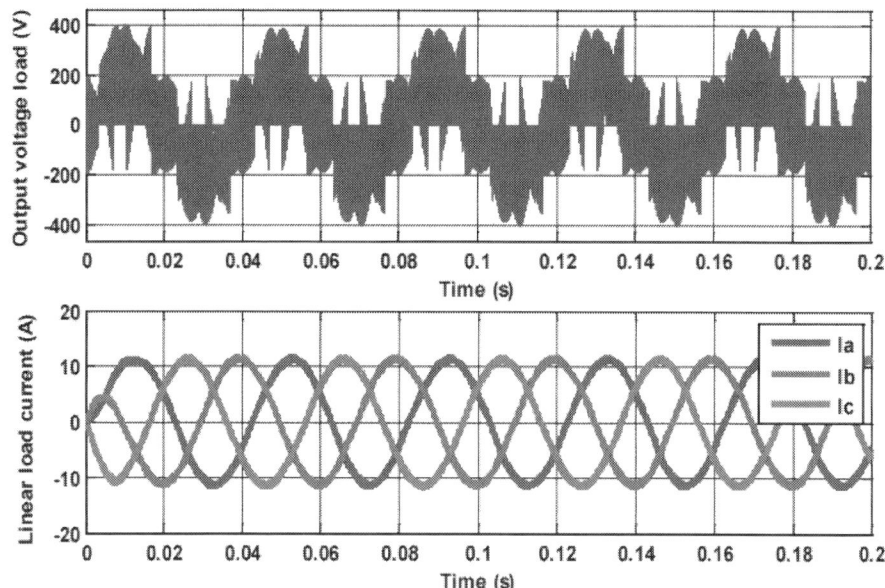

Figure 9. Output voltage and load current (PI, balanced grid and with constant I_{ref}).

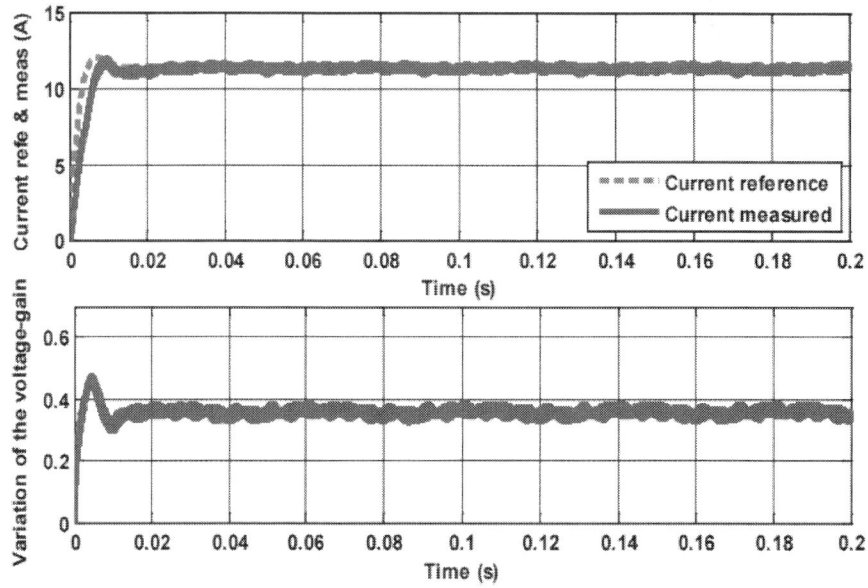

Figure 10. Load current and variation of the q (PI, balanced grid and with constant I_{ref}).

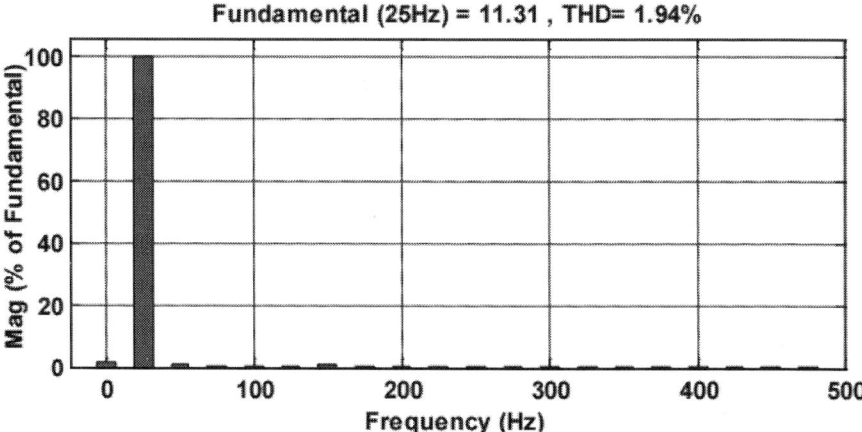

Figure 11. Harmonics spectrum of load current (PI, balanced grid and with constant I_{ref}).

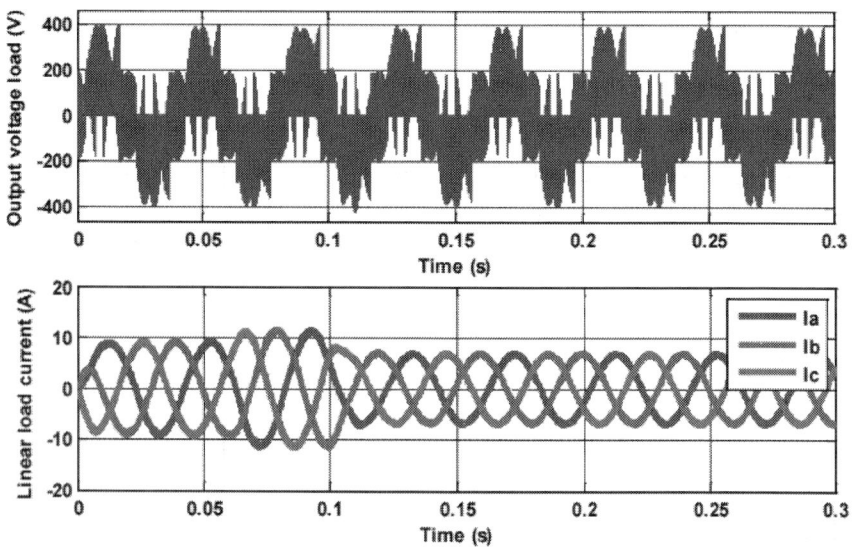

Figure 12. Output voltage and load current (PI, balanced grid and with stepped changing I_{ref}).

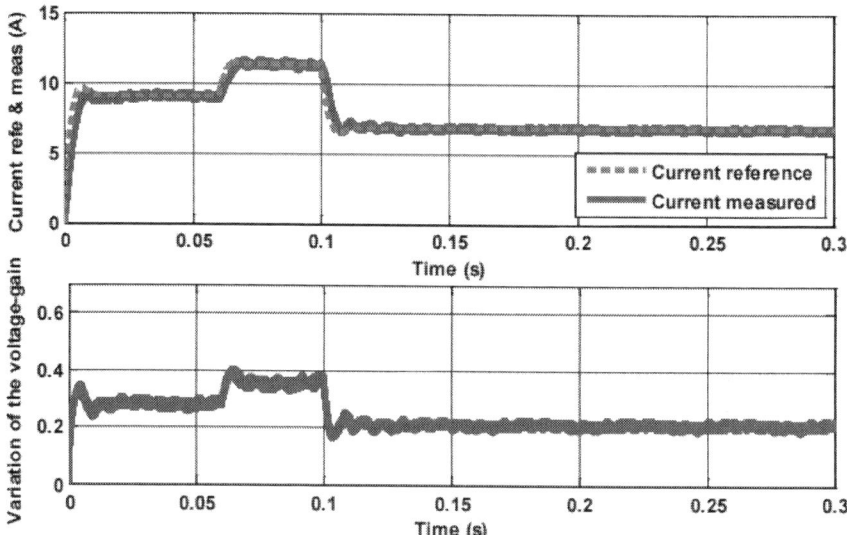

Figure 13. Load current and variation of the q (PI, balanced grid and with stepped changing I_{ref}).

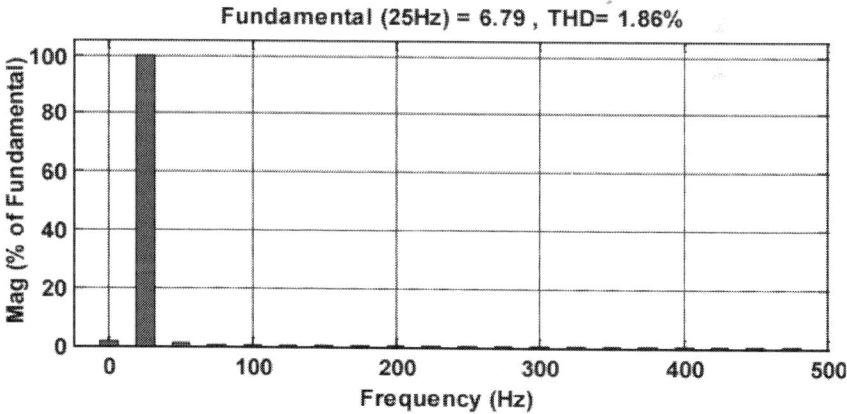

Figure 14. Harmonics spectrum of load current (PI, balanced grid an with stepped changing I_{ref}).

Balanced Grid Case with RST Controller

- Constant reference current I_{ref}:

Figure 15 shows the output voltage and linear load current using RST controller for balanced grid voltage with constant current reference. Figure 16 presents load current and variation of the ratio q. RST controller is used and the grid voltage balanced. Figure 17 shows the THD of load current with constant current reference.

- Time-varying reference current I_{ref}:

Figure 18 shows the output voltage and linear load current using RST controller for balanced grid voltage with stepped changing reference current. Figure 19 presents load current and variation of the ratio q. RST controller is used and the grid voltage balanced. Figure 20 shows the THD of load current with stepped changing reference current.

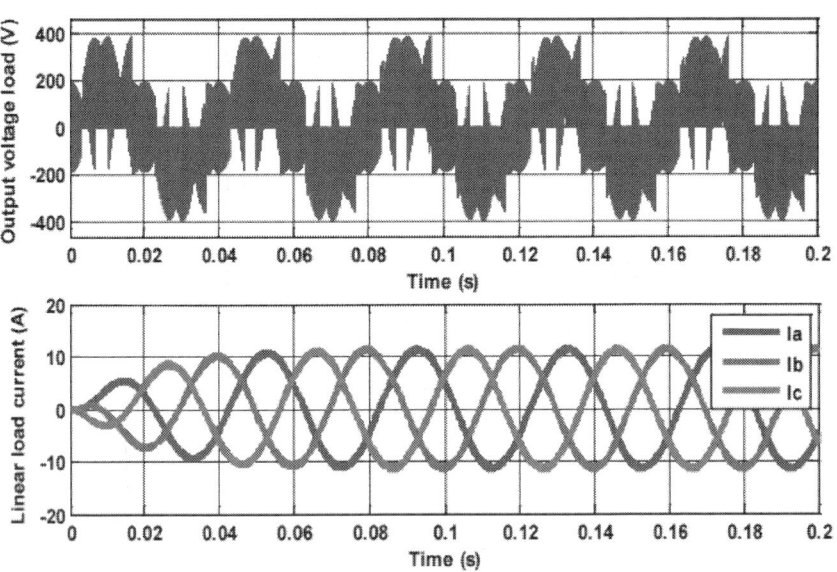

Figure 15. Output voltage and load current (RST, balanced grid and with constant I_{ref}).

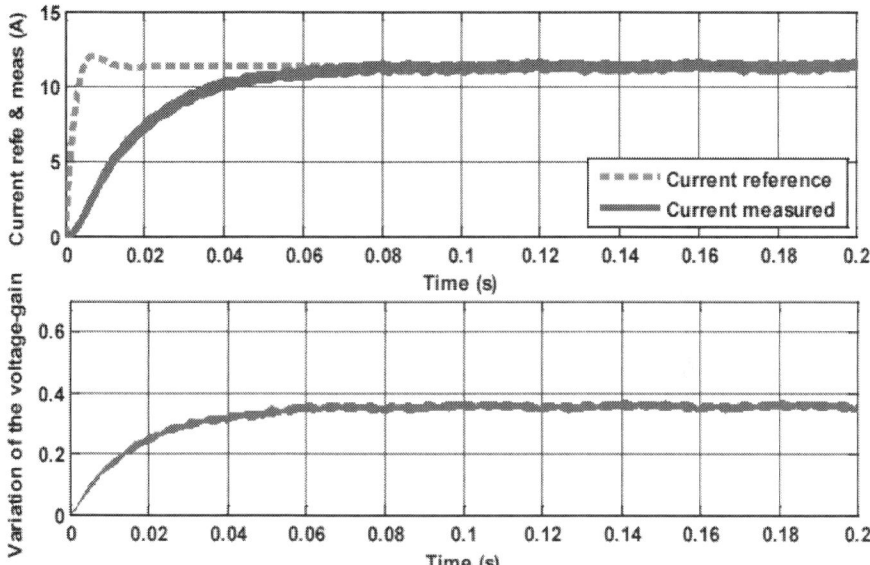

Figure 16. Load current and variation of the q (RST, balanced grid and with constant I_{ref}).

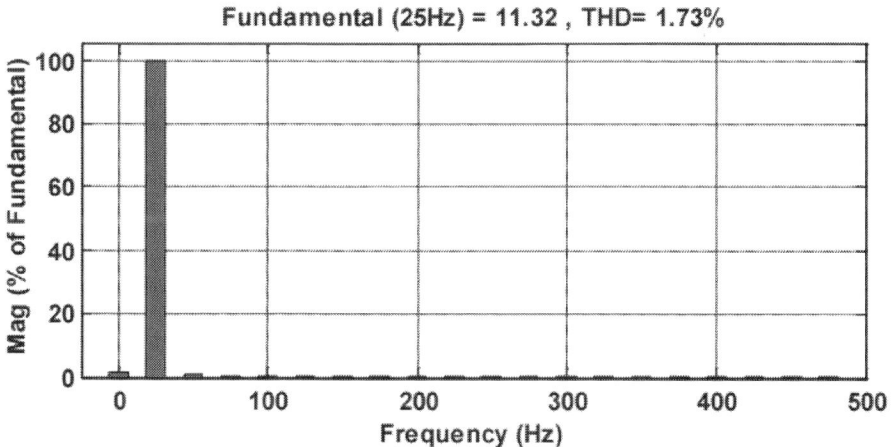

Figure 17. Harmonics spectrum of load current (RST, balanced grid and with constant I_{ref}).

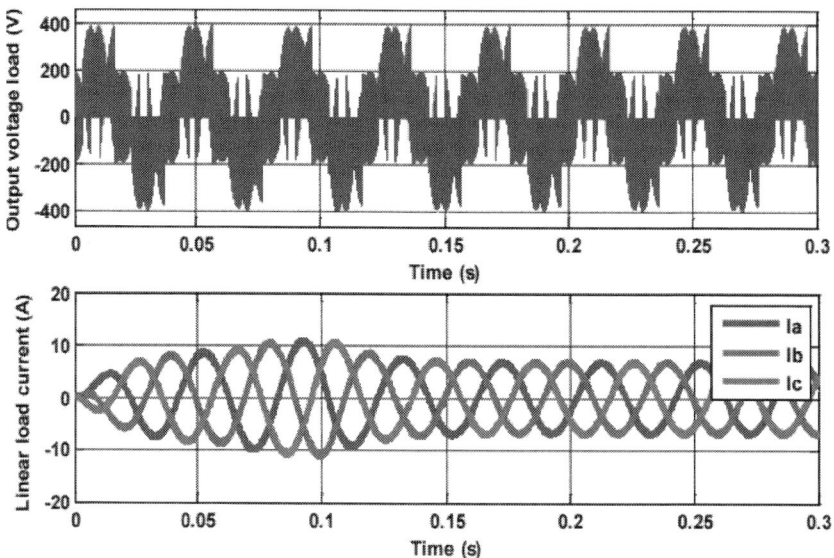

Figure 18. Output voltage and load current (RST, balanced grid and with stepped changing I$_{ref}$).

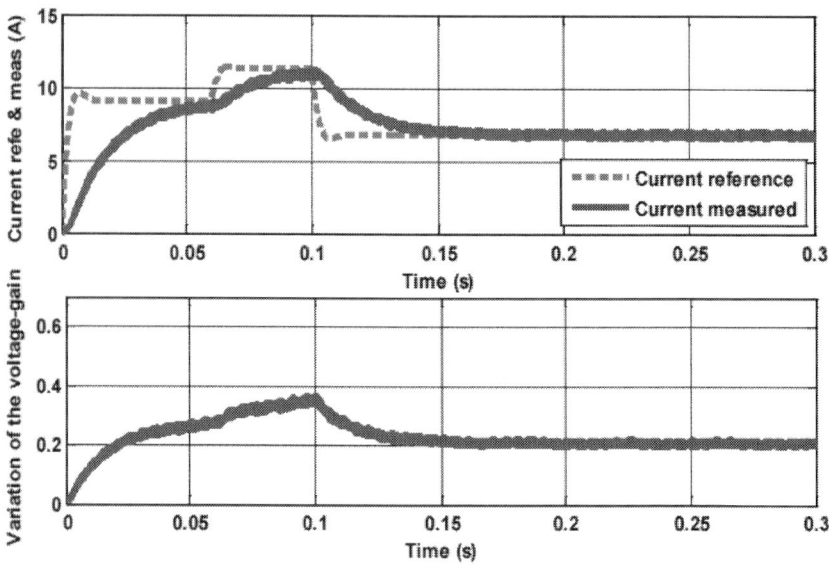

Figure 19. Load current and variation of the q (RST, balanced grid and with stepped changing I$_{ref}$).

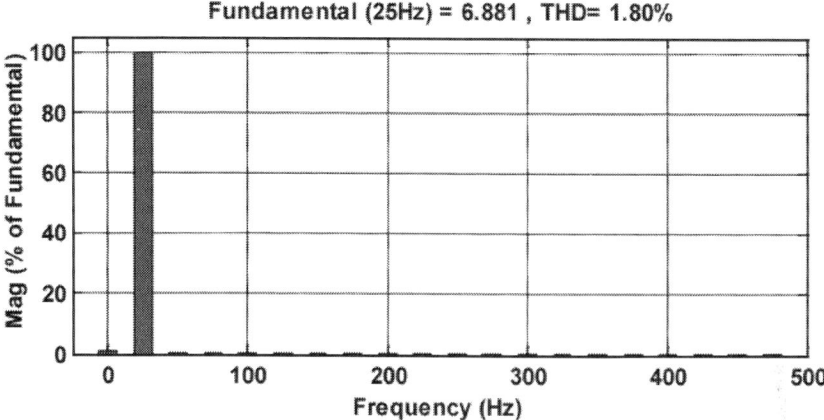

Figure 20. Harmonics spectrum of load current (RST, balanced grid an with stepped changing).

Unbalanced Grid Case with PI Controller

In this case, the amplitude of the input voltage of phase b is reduced to 20% relative to the phases a and c (Figure 21).

- Constant reference current I_{ref}:

Figure 22 shows the output voltage and linear load current using PI controller for unbalanced grid voltage with constant current reference. Figure 23 presents load current and variation of the ratio q. PI controller is used and the grid voltage unbalanced. Figure 24 shows the THD of load current with constant current reference.

- Time-varying reference current I_{ref}:

Figure 25 shows the output voltage and linear load current using PI controller for unbalanced grid voltage with stepped changing reference current. Figure 26 presents load current and variation of the ratio q. PI controller is used and the grid voltage unbalanced. Figure 27 shows the THD of load current with stepped changing reference current.

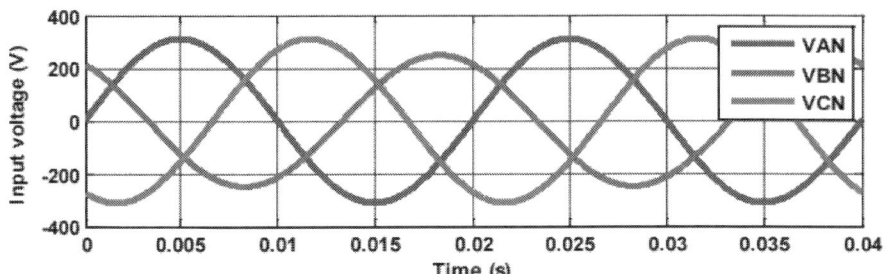

Figure 21. Unbalanced grid voltage.

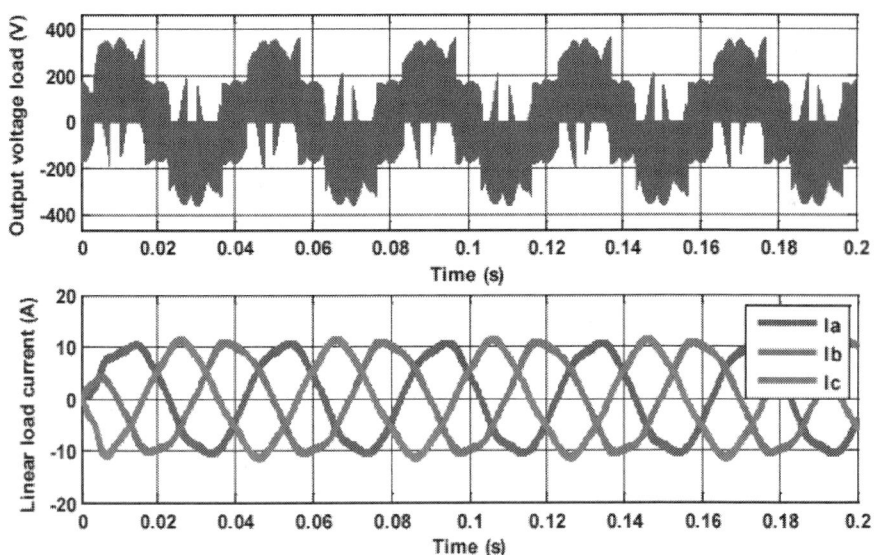

Figure 22. Output voltage and load current (PI, unbalanced grid and with constant I_{ref}).

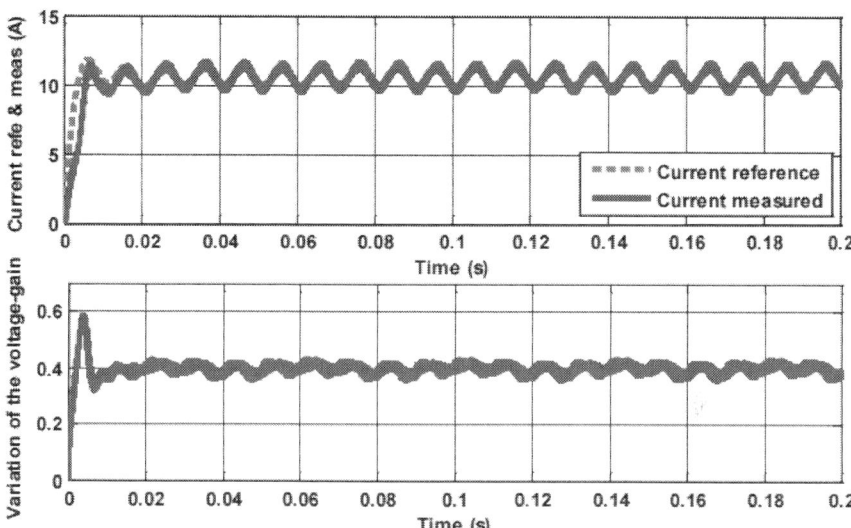

Figure 23. Load current and variation of the q (PI, unbalanced grid and with constant I_{ref}).

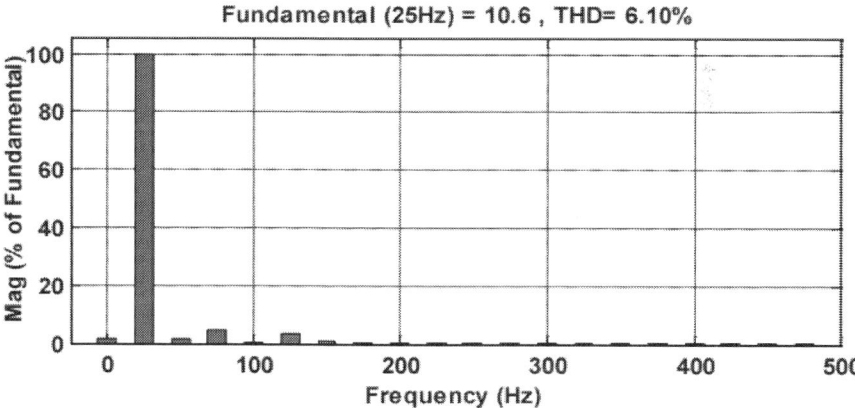

Figure 24. Harmonics spectrum of load current (PI, unbalanced grid and with constant I_{ref}).

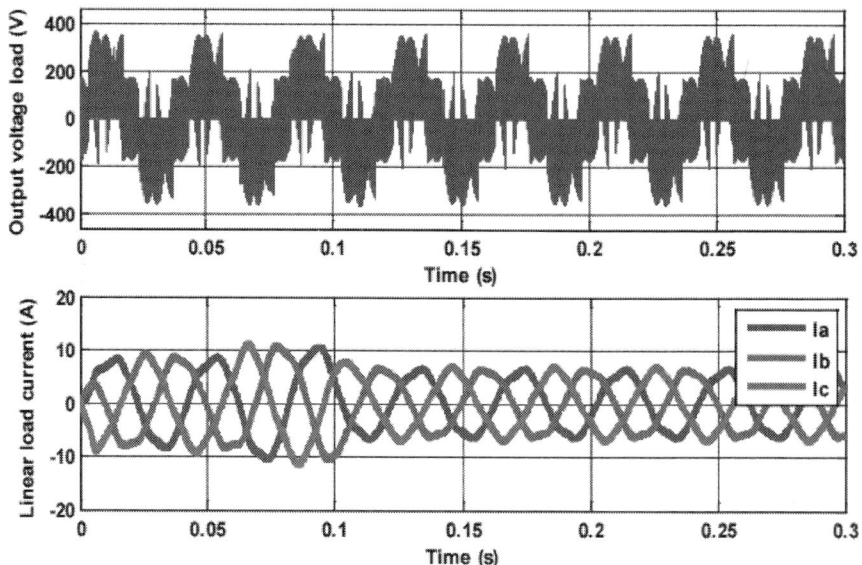

Figure 25. Output voltage and load current (PI, unbalanced grid and with stepped changing I_{ref}).

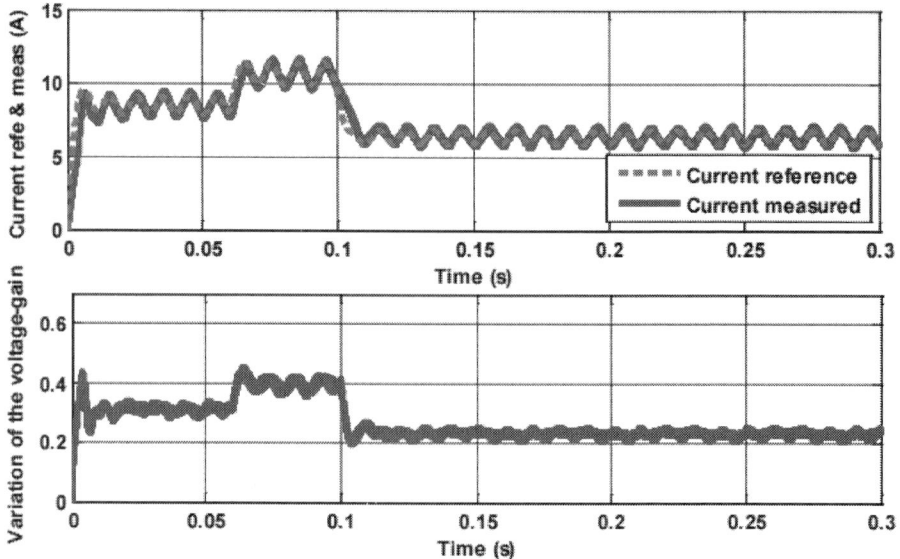

Figure 26. Load current and variation of the q (PI, unbalanced grid and with stepped changing I_{ref}).

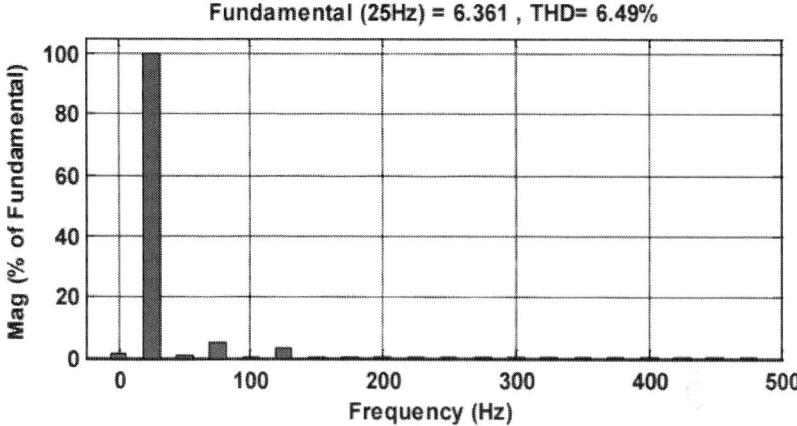

Figure 27. Harmonics spectrum of load current (PI, unbalanced grid an with stepped changing I_{ref}).

Unbalanced Grid Case with RST Controller

- Constant reference current I_{ref}:

Figure 28 shows the output voltage and linear load current using RST controller for unbalanced grid voltage with constant current reference. Figure 29 presents load current and variation of the ratio q. RST controller is used and the grid voltage unbalanced. Figure 30 shows the THD of load current with constant current reference.

- Time-varying reference current I_{ref}:

Figure 31 shows the output voltage and linear load current using RST controller for unbalanced grid voltage with stepped changing reference current. Figure 32 presents load current and variation of the ratio q. RST controller is used and the grid voltage unbalanced. Figure 33 shows the THD of load current with stepped changing reference current.

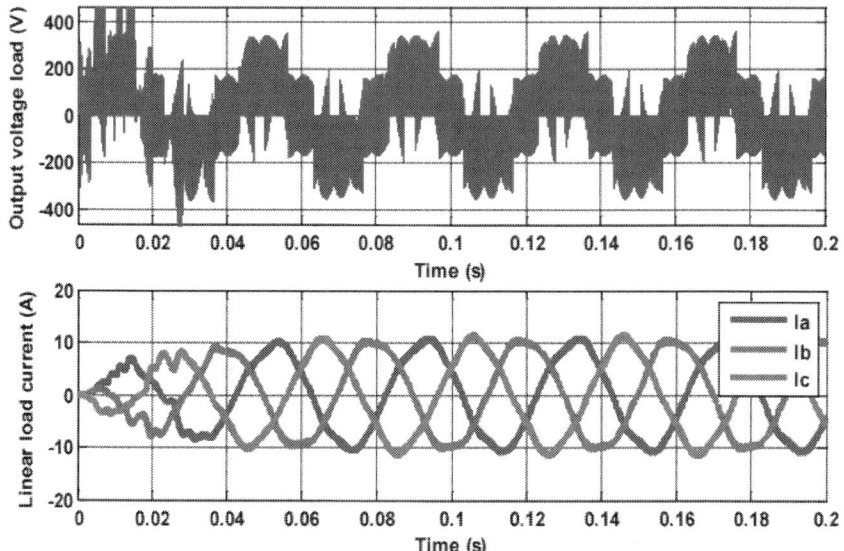

Figure 28. Output voltage and load current (RST, unbalanced grid and with constant I_{ref}).

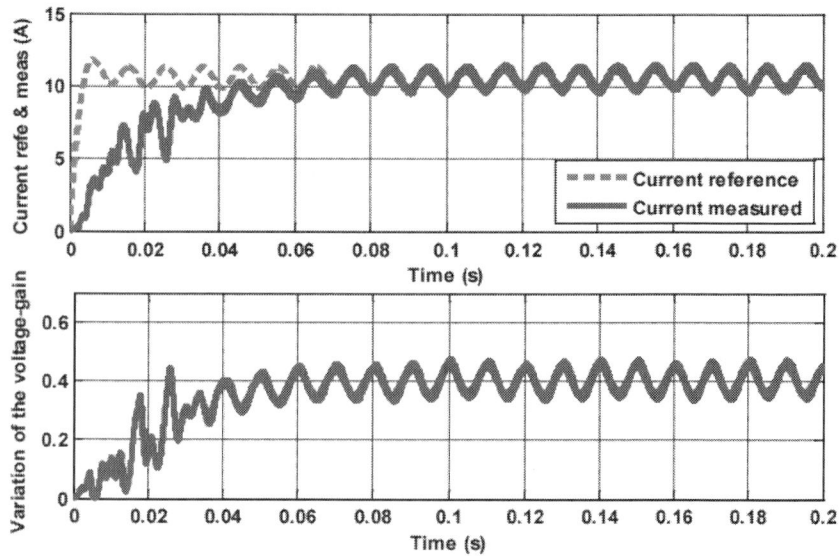

Figure 29. Load current and variation of the q (RST, unbalanced grid and with constant I_{ref}).

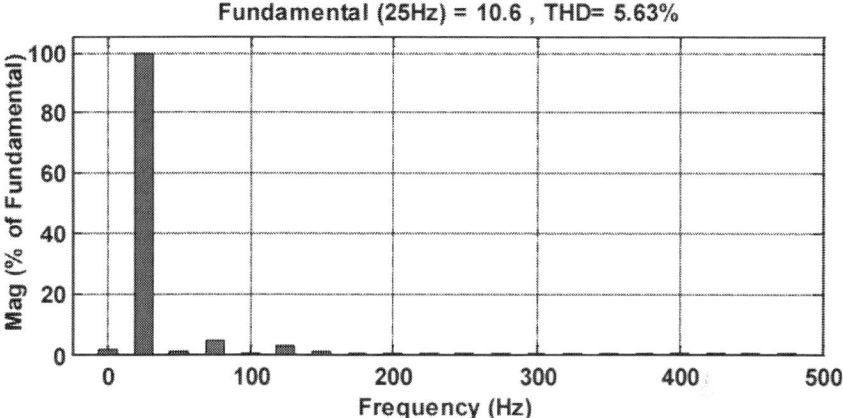

Figure 30. Harmonics spectrum of load current (RST, unbalanced grid and with constant I_{ref}).

Discussion the Results of Simulations

In Figure 9 and Figure 15, the voltage at the output of the matrix converter is formed by a succession of pulse widths conversely proportional to the frequency of the reference voltage $f_o = 25 \text{ Hz}$, and the RL load's current is almost sinusoidal with low Total Harmonic Distortion (THD) values. In Figure 22 and Figure 28, the voltage at the output of the matrix converter is formed by a succession of patterns which widths are proportional to the frequency of the reference voltage and the amplitude is $V_{im} = 300 \text{ V}$.

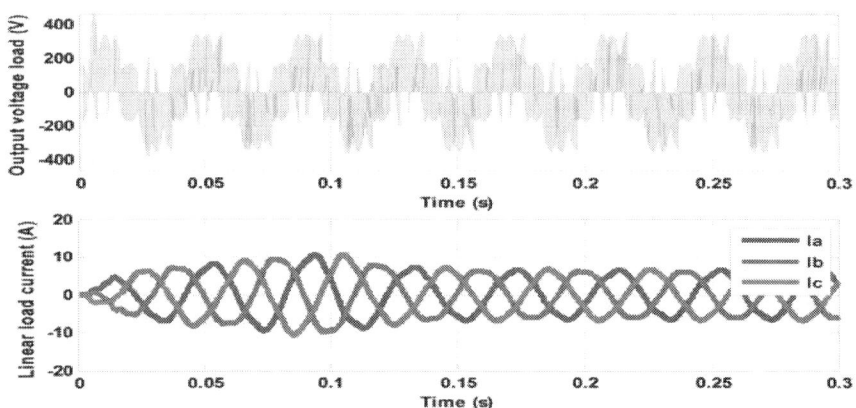

Figure 31. Output voltage and load current (RST, unbalanced grid and with stepped changing I_{ref}).

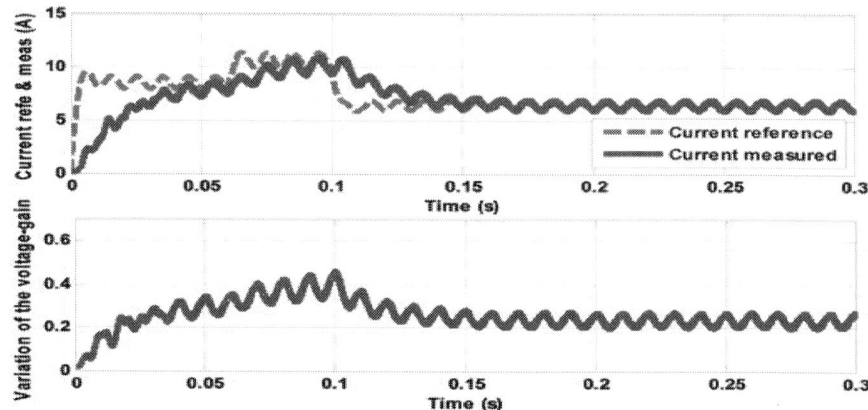

Figure 32. Load current and variation of the q (RST, unbalanced grid and with stepped changing I_{ref}).

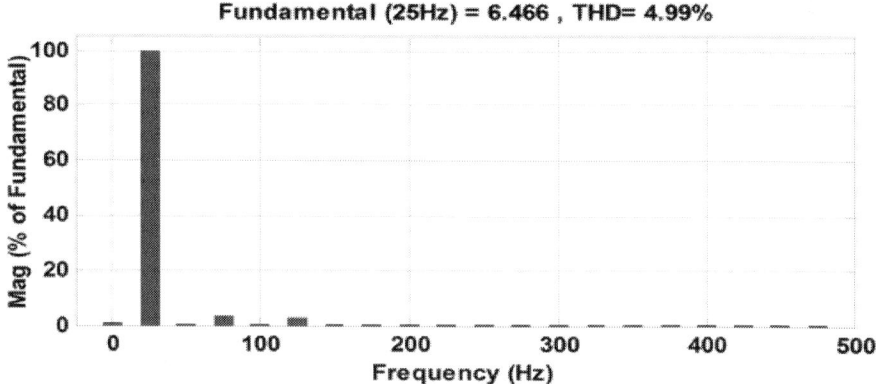

Figure 33. Harmonics spectrum of load current (RST, unbalanced grid an with stepped changing I_{ref}).

The THD increases for the unbalanced grid unlike in the balanced case (Figure 11 and Figure 17). However, the output currents are almost balanced, but are distorted. With the RST strategy, the signal quality of load current is much better than PI. Indeed, the THD is improved by 10.82% in the case of balanced grid, while this improvement is around 7.70% in the case of unbalanced grid Constant reference current I_{ref}. Note that in all the investigated cases, the gain q does not exceed 0.866.

Table 2 and Table 3 show the values of THD for balanced and unbalanced cases presented above.

Table 2. THD of load current with balanced grid.

Case with balanced grid	Values THD	IMP%
Constant I_{ref} (PI)	1.94%	10.82%
Constant I_{ref} (RST)	1.73%	
Time-varying I_{ref} (PI)	1.86%	3.220%
Time-varying of I_{ref} (RST)	1.80%	

Table 3. THD of load current with unbalanced grid.

Case with unbalanced grid	Values THD	IMP%
Constant I_{ref} (PI)	6.10%	7.700%
Constant I_{ref} (RST)	5.63%	
Time-varying I_{ref} (PI)	6.49%	23.11%
Time-varying of I_{ref} (RST)	4.99%	

Table 4. SSE with balanced grid.

Case with balanced grid	Values SSE
Constant I_{ref} (PI)	6.4827e + 04
Constant I_{ref} (RST)	1.2164e + 06
Time-varying I_{ref} (PI)	5.2185e + 04
Time-varying of I_{ref} (RST)	1.0137e + 06

Table 5. SSE with unbalanced grid.

Case with unbalanced grid	Values SSE
Constant I_{ref} (PI)	8.5052e+04
Constant I_{ref} (RST)	1.0992e+06
Time-varying I_{ref} (PI)	6.4179e+04
Time-varying of I_{ref} (RST)	9.4901e+05

Table 4 and Table 5 show the Sum Squared Error (SSE).

In terms of the response of the system and the static error, the PI controller gives little better results than RST controller as it can be seen the Table 4 and Table 5.

CONCLUSION

In this paper, a thorough theoretical modeling, analysis and comparison are presented for PI and RST control of MCs. A comprehensive control compensation method is used to find the PI gains. Moreover, the use of the pole placement technique is also shown to determine the RST's polynomial coefficients. Results for a balanced grid show lower load current THD as opposed to the unbalanced grid case, which is expected. However, RST control shows better performance. Nonlinear controllers tend to outperform these techniques at the expense of added complexity and computation. However, it is noteworthy that compared controllers are known for similar design complexity, which has been driving their use in the industry.

REFERENCES

1. Dendouga, A. (2010) Contrôle des puissances active et réactive de la machine asynchrone à double alimentation (DFIM). PhD Thesis, University of Batna, Batna.

2. Luis, F.P.A. (2011) Maximum Power Point Tracker of Wind Energy Generation Systems using Matrix Converters. Master's Thesis, Technical University of Lisbon, Lisbon.
3. Hulusi, K., Ramazan, A., Hüseyin, D., et al. (2008) A Novel Compensation Method Based on Fuzzy Logic Control for Matrix Converter under Distorted Input Voltage Conditions. Proceedings of the 2008 International Conference on Electrical Machines, Vilamoura, 6-9 September 2008, 1-5.
4. Venturini, M., Alesina, A., et al. (1980) The Generalised Transformer: A New Bidirectional Sinusoidal Waveform Frequency Converter with Continuously Adjustable Input Power Factor. Proceedings of the Power Electronics Specialists Conference (PESC' 80), Atlanta, 16-20 June 1980, 242-252.
5. Filho, M.E.O., Filho, E.R., Quindere, K.E.B., Gazoli, J.R., et al. (2006) A Simple Current Control for Matrix Con-verter. Proceedings of the International Symposium on Industrial Electronics, Montreal, 9-13 July 2006, 2090-2094.
6. Ram, G., Lincoln, S.A., et al. (2012) Fuzzy Adaptive PI Controller for Single Input Single Output Non-Linear System. ARPN Journal of Engineering and Applied, Sciences, 7, 1273-1280.
7. Hachicha, F., Krichen, L., et al. (2011) Performance Analysis of a Wind Energy Conversion System Based on a Doubly-Fed Induction Generator. Proceedings of the 8th International Multi-Conference on Systems, Signals & Devices, Sousse, 22-25 March 2011, 1-6.
8. Bouhamida, M., Denai, M.A., et al. (2005) Robust Stabilizer of Electric Power Generator Using H∞ with Placement Constraints. Journal of Electrical Engineering, 56, 176-182.
9. Oubelli, A.L. (2011) Mise En æuvre d'un modèle générique du convertisseur matriciel dans les environnements EMTP-RV et MATLAB-SIMULINK. Master's thesis, Ecole Polytechnique de Montréal, Montréal.
10. Hamane, B., Doumbia, M.L., Cheriti, A., Belmokhtar, K., et al. (2014) Comparative Analysis of PI and Fuzzy Logic Controllers for Matrix Converter. Proceedings of the 9th International Conference on Ecological Vehicles and Renewable Energies (EVER), Monte-Carlo, 25-27 March 2014, 25-27.
11. Hamane, B., Doumbia, M.L., Cheriti, A., Belmokhtar, K., et al. (2013) Modeling and Control of a Matrix Converter Using Fuzzy Supervisory Controller. Proceedings of the 3rd International Conference on Systems and Control (ICSC), Algiers, 29-31 October 2013, 433-438.
12. Afonso, L.P. (2011) Maximum Power Point Tracker of Wind Energy Generation Systems Using Matrix Converters. Master's Thesis, Higher Technical Institue of Technical University of Lisbon, Lisbon.

13. Boukadoum, A., Bahi, T., Oudina, S., Souf, Y., Lekhchine, A.S., et al. (2012) Fuzzy Control Adaptive of a Matrix Converter for Harmonic Compensation Caused by Nonlinear Loads. Energy Procedia, 18, 715-723.
14. Ghedamsi, K. (2008) Contribution à la modélisation et la commande d'un convertisseur direct de fréquence Application à la conduite de la machine asynchrone. PhD Thesis, National Polytechnic School of Process Control Laboratory, El-Harrach.
15. Dendouga, A., Abdessemed, R., Essounbouli, N., Megherbi, A.C., et al. (2013) Robustness Evaluation of Vector Control of Induction Motor fed by SVM Matrix Converter. 3rd International Conference on Systems and Control (ICSC), Algiers, 165-170.
16. Rodriguez, S.E., Blaabjerk, F., et al. (1985) Modelling, Analysis and Simulation of Matrix Converters. Applications, IA-21, 1337-1342.
17. Belabbes, A., Hamane, B., Bouhamida, M., Draou, A., Benghanem, M., et al. (2012) Power Control of a Wind Energy Conversion System based on a Doubly Fed Induction Generator using RST and Sliding Mode Controllers. Proceedings of the International Conference on Renewable Energies and Power Quality (ICREPQ'12), Santiago de Compostella, 28-30 March 2012. http://www.icrepq.com/icrepq%2712/298-belabbes.pdf
18. Mai, T.D., Mai, B.L., Pham, D.T., Nguyen, H.P., et al. (2007) Control of Doubly-Fed Induction Generators Using Dspace R&D Controller Board—An Application of Rapid Control Coordinated with Matlab/Simulink. Proceedings of the International Symposium on Electrical & Electronics Engineering, 3, 302-307.

CITATION

Bekhada Hamane, Mamadou, Lamine Doumbia, Hicham Chaoui, Mohamed Bouhamida, AhmedChériti, Mustapha Benghanem, (2015) PI and RST Control Design and Comparison for Matrix Converters Using Venturini Modulation Strategy. *Journal of Power and Energy Engineering*, **03**, 36-54. doi: 10.4236/jpee.2015.38005

CHAPTER 7

A Systematic Method for Designing a PR Controller and Active Damping of the LCL Filter for Single-Phase Grid-Connected PV Inverters

Ningyun Zhang, Houjun Tang and Chen Yao

School of Electronic Information and Electrical Engineering, Shanghai Jiao Tong University, 800 Dongchuan Rd, Minhang District, Shanghai 200240, China;

ABSTRACT

The Proportional Resonant (PR) current controller provides gains at a certain frequency (resonant frequency) and eliminates steady state errors. Therefore, the PR controller can be successfully applied to single grid-connected PV inverter current control. On the contrary, a PI controller has steady-state errors and limited disturbance rejection capability. Compared with the L- and LC filters, the LCL filter has excellent harmonic suppression capability, but the inherent resonant peak of the LCL filter may introduce instability in the whole system. Therefore, damping must be introduced to improve the control of the system. Considering the controller and the LCL filter active damping as a whole system makes the controller design method more complex. In fact, their frequency responses may affect each other. The traditional trial-and-error procedure is too time-consuming and the design process is inefficient. This paper provides a detailed analysis of the frequency response influence between the PR controller and the LCL filter regarded as a whole system. In addition, the paper presents a systematic method for designing controller parameters and the capacitor current feedback coefficient factor of LCL filter active-damping. The new method relies on meeting the stable margins of the system. Moreover, the paper also clarifies the impact of the grid on the inverter output current. Numerical simulation and a 3 kW laboratory setup assessed the feasibility and effectiveness of the proposed method.

INTRODUCTION

The rapid increase in global energy consumption has accelerated the need for greener energy sources. Nowadays, renewable, highly sustainable energies derived from inexhaustible sources such as wind, photovoltaic, or tides have attracted much more attention [1]. Distributed power generation systems (DPGS) are attractive both for the market and for researchers, and the most important part of a DPGS, the grid-connected inverter, is a research hotspot.

The filter is an essential component that suppresses the harmonics introduced through the Pulse-Width Modulation (PWM) technique used in grid-connected inverters. Thanks to the excellent harmonic suppression ability an LCL filter topology is much more attractive than L and LC filters. However, the LCL filter generates significant resonance peaks that worsen the system stability, and the control scheme design of grid-connected inverters take into consideration this behavior. Damping the filter resonance overcomes LCL filter oscillations and stabilizes the system [2,3]. Simple passive damping with a resistor connected in series or parallel to the LCL filter capacitor, results in additional power loss and decreases the LCL filter performance [4]. Papers in the literature present different active damping methods. The so-called 'active damping by well-designed control algorithm' method is usually preferred because it has no additional power losses [4,5]. Among the active damping methods, the capacitor current feedback control algorithm is important because it is simple to manipulate and is stable [6].

The quality of the injected grid current is important in grid-connected inverter control. Because of the infinite gain of the PR controller at the selected resonant frequency, the zero steady-state error can be achieved [7,8]. Papers in the literature do not discuss in detail the analysis of the frequency response influence between the PR controller and LCL filter. Moreover, the design methods of PR controllers and active damping of the LCL filter have not been well clarified. In most cases, many trial-and-error procedures have been carried out to obtain a set of parameters.

Considering the controller and the LCL filter active damping as a whole system enhances the complexity of controller design method. Their frequency responses may influence each other and affect the system stability. The design of the PI controller and the LCL filter active damping has been thoroughly investigated in [6]. However, a systematic study of the design procedures of the PR controller and the LCL filter active damping is missing. This paper discusses in detail such a design method. When adopting the PR controller instead of PI controller, the grid may produce a different impact on the inverter output current. The paper also deals with the grid impact and proposes a method to eliminate the grid effect on the output currents of the inverter.

SYSTEM OVERVIEW AND NUMERICAL MODELING

Figure 1 shows the typical topology of a two-stage single-phase grid-connected photovoltaic (PV) system.

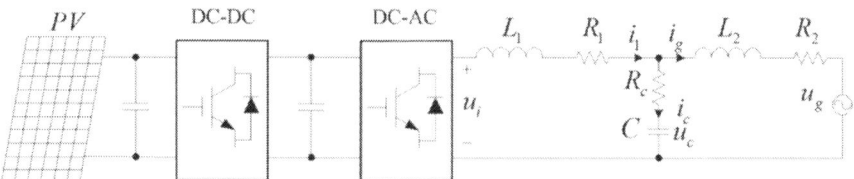

Figure 1. Two-stage single-phase grid-connected PV system with LCL filter typical topology.

The series connected $L_1 + R_1$, $L_2 + R_2$ and $C + Rc$ that compose the LCL-type filter attenuate the harmonic injected into the grid generated from the inverter with the PWM technique. The boost DC-DC converter connected to the photovoltaic panel step up the voltage of the DC bus to a proper level for the DC-AC inverter. The H-bridge DC-AC inverter produces proper sinusoidal current in the grid with unity power factor based on the Maximum Power Point Tracking (MPPT) and Phase Locked Loop (PLL) algorithms. According to Figure 1, the mathematical model of the grid-connected inverter and the LCL filter is:

$$\begin{cases} L_1 \frac{di_1}{dt} + R_1 i_1 = u_i - u_c - R_c i_c \\ L_2 \frac{di_g}{dt} + R_2 i_g = u_c - u_g - R_c i_c \\ C \frac{du_c}{dt} = i_c \\ i_1 = i_g + i_c \end{cases} \tag{1}$$

CONTROL SCHEME

Figure 2 shows the PV system control scheme. By sampling the PV panel current and the voltage a proper MPPT algorithm can be ran. Papers in the literature present different MPPT algorithms [9–12]. The DC bus voltage control algorithm gives the injected current reference. Moreover, a PLL algorithm synchronizes the injected current with the grid voltage. Wide discussions on PLL algorithms for single-phase inverters can be found in various papers in the literature [13,14].

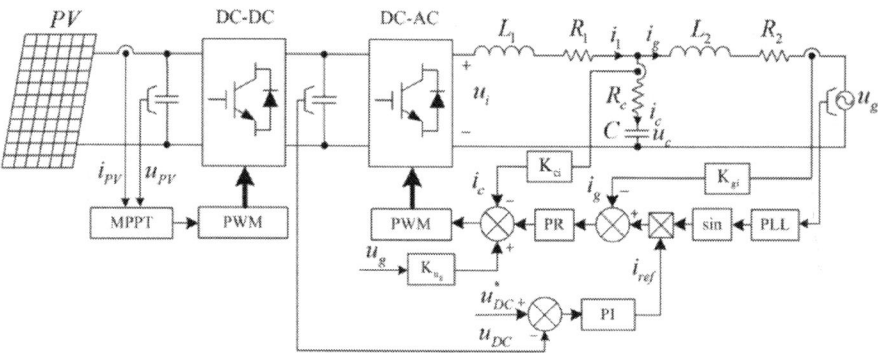

Figure 2. Two-stage single-phase PV system with LCL-filter control scheme.

The current reference is sinusoidal: since a PI controller is unable to track a sinusoidal reference without steady-state errors, the PR controller has been adopted. The PR controller tracks the current introducing an infinite gain at a certain frequency (resonant frequency) [8,15]. Sampling the LCL filter capacitor current modifies the control algorithm to perform an active damping; the active damping is mostly used to smooth the resonance peak of the LCL filter. The feedforward of grid voltage is essential and the following sections of the paper will focus on it.

According to Equation (1) and Figure 2, Figure 3 shows the control diagram of injected current. To simplify the analysis, resistors R_1, R_2, and R_c have been omitted: this corresponds to the worst LCL filter working condition. Figure 4 shows the equivalent control block diagram.

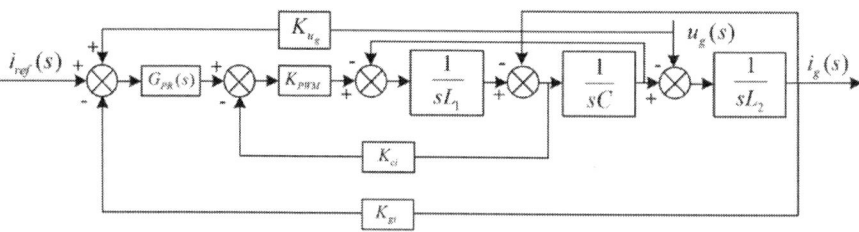

Figure 3. Injected current: control diagram.

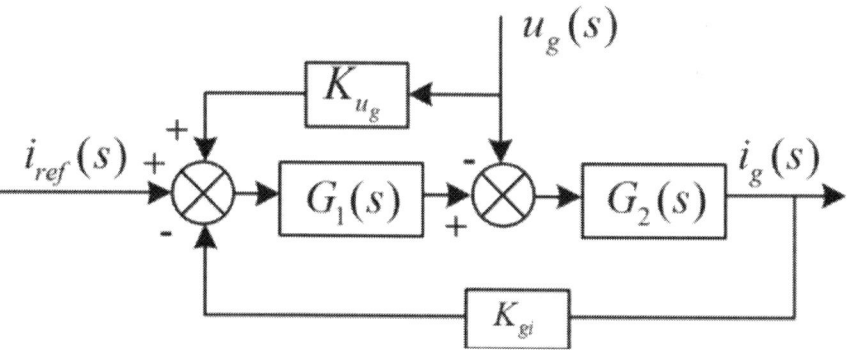

Figure 4. Equivalent control diagram.

where:

$$G_1(s) = \frac{G_{PR}(s)K_{PWM}}{s^2 L_1 C + sCK_{PWM}K_{ci} + 1} \tag{2}$$

$$G_2(s) = \frac{s^2 L_1 C + sCK_{PWM}K_{ci} + 1}{s^3 L_1 L_2 C + s^2 L_2 CK_{PWM}K_{ci} + s(L_1 + L_2)} \tag{3}$$

PR Controller

The Laplace transform of the ideal PR controller is:

$$G_{PR}(s) = K_P + \frac{2K_r s}{s^2 + \omega_1^2}$$

where K_p is the proportional gain and ω_1, K_r are the resonant frequency and gain, respectively. The PI controller provides an infinite gain with a constant variable; it get a quick response to a step reference without steady-state error, but is unable to track a sinusoidal reference. On the contrary, the PR controller provides an infinite gain at the selected frequency (resonant frequency) and zero phase-shift. Figure 5shows the Bode diagram of an ideal PR controller. The controller cannot be realized in a physical circuit since it is lossless [8], and the improved form of the controller is a practical alternative.

Equation (4) shows the PR controller improved form; the integral term has been approximated using a high-gain low-pass filter [8]:

$$G_{PR}(s) = K_P + \frac{2K_r \omega_{PRc} s}{s^2 + 2\omega_{PRc} s + \omega_1^2} \tag{4}$$

where ω_{PRc} is the bandwidth at -3 dB cutoff frequency of the controller that reduces the sensitivity of the grid fundamental frequency variation. The gain of the controller at $\omega_1 - \omega_{PRc}$ and $\omega_1 + \omega_{PRc}$ is $2\sqrt{}$ [8].Figure 6 shows the Bode diagram of the improved PR controller. The proportional gain K_p mainly determines the dynamics of the controller, while K_r determines the amplitude gain at a selected frequency, and controls the bandwidth around it.

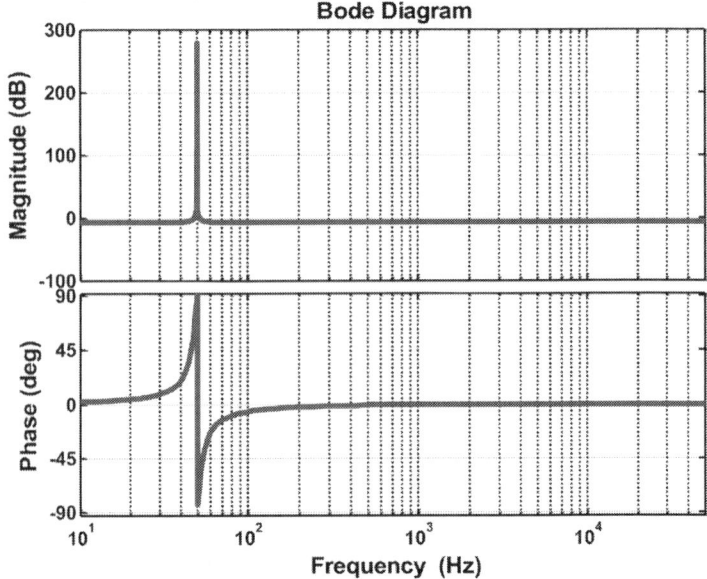

Figure 5. Ideal PR controller Bode diagram.

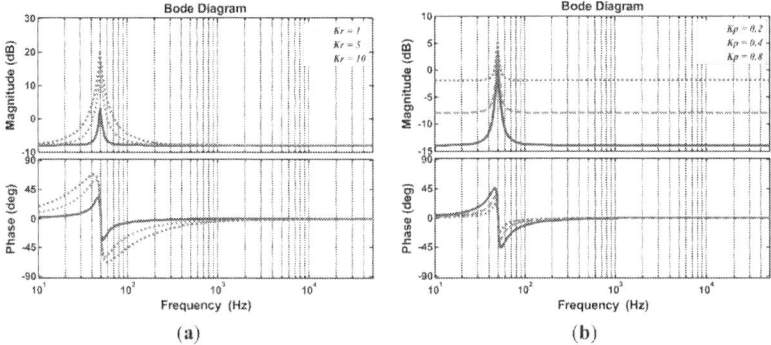

Figure 6. Improved PR controller Bode diagram: (a) $K_p = 0.4$; (b) $K_r = 1$.

LCL Filter

LCL filter is a high order system; the Laplace transfer function is given by following Equation (R_1, R_2, and R_c have been omitted):

$$G_{LCL}(s) = \frac{1}{s^3 L_1 L_2 C + s(L_1 + L_2)}$$

The LCL filter resonant frequency is:

$$f_{res} = \frac{1}{2\pi}\sqrt{\frac{L_1 + L_2}{L_1 L_2 C}}$$

The LCL-filter provides outstanding performance attenuating the switching frequency current harmonics, better than the L- and LC-filters. Figure 7 compares the Bode diagram of the LCL and L filters. The red curve LCL PD (passive damping) represents the characteristic of the LCL-filter with a resistor connected in series with the capacitor. At low frequencies range, LCL and L filters show similar dynamic behaviors. At the high frequency range, LCL filter has stronger attenuation ability than the L-filter; this feature applies to medium and large power applications, since the very low switch frequency. As a drawbacks, the LCL filter presents resonance peaks around the resonant frequency and the phase-frequency curve across $-\pi$. As a result, systems become highly sensitive to disturbances and unstable, and the control algorithm design must consider the presence of an effective damping method.

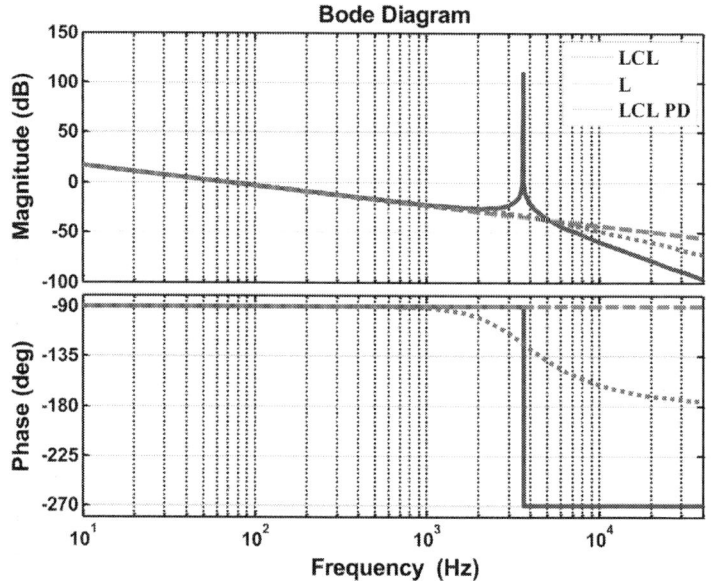

Figure 7. LCL, and L filters Bode diagrams.

The red curve LCL PD shows that the damping smooths the resonance peak, but worsens the LCL filter performance at the high frequency range, and generates additional power losses. Thus, an active damping method that modifies the control algorithm is preferable. This study adopts the active damping method based on the capacitor current feedback; the detailed design process is discussed later in the paper.

Grid Impact and Feedforward Control

With reference to Figure 4, the current injected into the grid can be written as:

$$i_g(s) = \frac{1}{K_{gi}} \frac{G(s)}{1+G(s)} i_{ref}(s) - \frac{1-K_{u_g}(s)G_1(s)}{1+G(s)} G_2(s)u_g(s) = i_{g1}(s) + i_{g2}(s) \qquad (5)$$

where:

$$G(s) = G_1(s)G_2(s)K_{gi},$$

$$i_{g1}(s) = \frac{1}{K_{gi}} \frac{G(s)}{1+G(s)} i_{ref}(s), \text{and}$$

$$i_{g2}(s) = -\frac{1-K_{ug}(s)G_1(s)}{1+G(s)} G_2(s)u_g(s)$$

Defining the ratio between $i_{g2}(s)$ and $i_{g1}(s)$ as:

$$\sigma(s) = \frac{i_{g2}(s)}{i_{g1}(s)} = -\frac{[1 - K_{ug}(s)G_1(s)]G_2(s)u_g(s)K_{gi}}{G(s)i_{ref}(s)} \tag{6}$$

The grid can be treated as a disturb; omitting the feedforward from the grid the Equation (6) becomes:

$$\sigma(s) = \frac{i_{g2}(s)}{i_{g1}(s)} = -\frac{u_g(s)}{G_1(s)i_{ref}(s)} \tag{7}$$

Taking the grid fundamental frequency as subject, the capacitor in LCL filter can be omitted since the resonant frequency of LCL filter is far greater than the fundamental frequency [6]. Thus, Equation (4) can be written as Equation (8):

$$G_{PR}(j\omega_1) = K_p + \frac{2K_r\omega_{PRc}j\omega_1}{(j\omega_1)^2 + 2\omega_{PRc}j\omega_1 + \omega_1^2} = K_p + K_r \tag{8}$$

Substituting Equations (7) and (8) into Equation (6), gives:

$$\sigma(j\omega_1) = -Z_g(j\omega_1)\frac{1}{K_{PWM}(K_p + K_r)} \tag{9}$$

where: $Z_g(j\omega_1) = \frac{u_g(s)}{i_{ref}(s)} \cdot Z_g(j\omega_1)$ is the grid impedance at the fundamental frequency ω_1. Because of the unity power factor, $Z_g(j\omega_1)$ is a pure resistor. Equation (9) shows the grid impact on the current: the grid-generated current i_{g2} flows in the opposite direction of the current i_{g1}. It is very different from a system that adopts PI controller as current controller [6] in which the grid generates a $\pi2$ lag component to i_{g1}. According to Equations (5) and (9), the phase diagram of the grid can be drawn as Figure 8 shown:

Therefore, the feedforward of the grid is required; setting i_{g2} to zero, the feedforward factor becomes:

$$K_{u_g}(j\omega_1) = \frac{1}{G_1(j\omega_1)} = \frac{1}{K_{PWM}(K_p + K_r)} \tag{10}$$

If the feedforward control lies behind the PR controller, the feedforward factor has a very simple form:

$$K_{u_g}(j\omega_1) = \frac{1}{G_1(j\omega_1)} = \frac{1}{K_{PWM}} \tag{11}$$

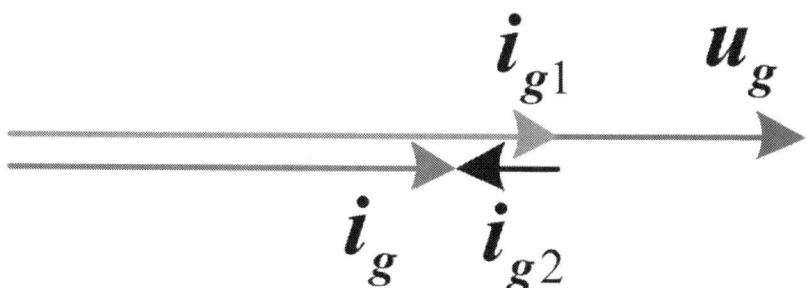

Figure 8. Grid phase diagram.

PR CONTROLLER AND ACTIVE-DAMPING OF LCL FILTER DESIGN

PR Controller Design
Proportional Gain K_p
Assuming that the fundamental frequency of the grid varies in the range ±1 Hz, ω_{PRc} is:

$$\omega_{PRc}=2\pi \tag{12}$$

Following Equation describes the control diagram open-loop transfer function:

$$G(s)=G1(s)G2(s)K_{gi} \tag{13}$$

The relationship between the cutoff frequency f_c, the sampling frequency f_s, and the resonant frequency f_{res} is [6,16]:

$$f_c < \tfrac{1}{10}f_s, \quad \text{and} \quad \tfrac{1}{4}f_s < f_{res} < \tfrac{1}{2}f_s$$

]The cutoff frequency f_c of the system is usually designed to be far lower than the sampling frequency f_s, and much smaller than the resonant frequency f_{res} of LCL filter. Therefore, considering the frequency response of the system lower than the cutoff frequency, the capacitor of LCL filter can be omitted [6], andEquation (13) can be rewritten as Equation (14):

$$G(s) \approx \frac{K_{gi} K_{PWM} G_{PR}(s)}{s(L_1 + L_2)} \tag{14}$$

The cutoff frequency f_c is higher than the fundamental frequency f_1. As a result, according to Figure 6the PR controller expression simplifies. At cutoff frequency, the magnitude frequency response of the system is zero, and Equation (15) is obtained:

$$20 \lg |G(j2\pi f_c)| \approx 20 \lg \left| \frac{K_{gi} K_{PWM} K_p}{j2\pi f_c (L_1 + L_2)} \right| = 0 \tag{15}$$

By successive approximations of Equation (15), the gain K_p is:

$$K_p \approx \frac{2\pi f_c (L_1 + L_2)}{K_{gi} K_{PWM}} \tag{16}$$

Fundamental Frequency Gain K_r

The tracking steady-state error of the grid current i_g can be calculated as:

$$E_g(s) = \frac{K_{gi}[i_{g1}(s) + i_{g2}(s)] - i_{ref}(s)}{i_{ref}(s)} = \underbrace{\frac{G(s)}{1 + G(s)}}_{-1} - \frac{[1 - K_{u_g}(s)G_1(s)]K_{gi}G_2(s)Z_g(s)}{1 + G(s)} \tag{17}$$

Assuming the adoption of grid feedforward control, Equation (17) can be simplified as:

$$E_g(s) = \frac{K_{gi} i_{g1}(s) - i_{ref}(s)}{i_{ref}(s)} = \frac{G(s)}{1 + G(s)} - 1 \tag{18}$$

Considering the fundamental frequency f_1, and supposing the steady-state error of i_g is η, following Equation gives the minimum of the magnitude-frequency response of open-loop transfer function at the fundamental frequency f_1:

$$|G_{\min}(j2\pi f_1)| = \frac{1}{\eta} - 1$$

That is:

$$|G(j2\pi f_1)| \approx \left| \frac{K_{gi} K_{PWM} G_{PR}(j2\pi f_1)}{j2\pi f_1 (L_1 + L_2)} \right| \geq |G_{\min}(j2\pi f_1)| \tag{19}$$

Taking Equation (16) into Equation (19), the lower threshold of K_r is deduced as:

$$K_r \geq \frac{(2\pi f_1 |G_{\min}(j2\pi f_1)| - 2\pi f_c)(L_1 + L_2)}{K_{gi} K_{PWM}} \quad (20)$$

The phase-frequency response of PR controller can be written as:

$$\phi(\omega) = \arctan \frac{\omega_1^2 - \omega_c^2}{2\omega_{PRc}\omega_c}\left(1 + \frac{K_r}{K_p}\right) - \arctan \frac{\omega_1^2 - \omega_c^2}{2\omega_{PRc}\omega_c} \quad (21)$$

As the improved PR controller Bode plot shows (Figure 6), the larger K, the larger gain at the fundamental frequency is. It is beneficial to the quick response speed, but it emphasizes an extra drawback. The phase response of the controller tends to as K_r increases, which would affect the phase margins of the whole PV system. Therefore, a compromise between the response speed and the stability is required. Assuming the phase-frequency response of PR controller at cutoff frequency not exceeds $-\phi$, and that the phase response of the other components at cutoff frequency must not exceed $(PM - \pi + \phi)$, following Equation (22) describes the phase:

$$\varphi(\omega_c) = \arctan \frac{2\omega_{PRc}\omega_c}{\omega_1^2 - \omega_c^2}\left(1 + \frac{K_r}{K_p}\right) - \arctan \frac{2\omega_{PRc}\omega_c}{\omega_1^2 - \omega_c^2} \geq -\phi \quad (22)$$

Taking Equation (16) into Equation (22), the upper threshold of K_r is:

$$K_r \leq \frac{\omega_c(L_1 + L_2)}{K_{gi}K_{PWM}}\left[\frac{\omega_1^2 - \omega_c^2}{2\omega_{PRc}\omega_c}\tan\left(\arctan \frac{2\omega_{PRc}\omega_c}{\omega_1^2 - \omega_c^2} - \phi\right) - 1\right] \quad (23)$$

According to Equations (20) and (23), the range of K_r is defined through Equation (24):

$$\frac{(\omega_1 |G_{\min}(j\omega_1)| - \omega_c)(L_1 + L_2)}{K_{gi}K_{PWM}} \leq K_r \leq \frac{\omega_c(L_1 + L_2)}{K_{gi}K_{PWM}}\left[\frac{\omega_1^2 - \omega_c^2}{2\omega_{PRc}\omega_c}\tan\left(\arctan \frac{2\omega_{PRc}\omega_c}{\omega_1^2 - \omega_c^2} - \phi\right) - 1\right] \quad (24)$$

Active-Damping of LCL Filter

Figure 9 shows the Bode diagram of the PV system with no PR controller. A high value of K_{ci} shows better resonance peak damping capability, but phase margins become smaller. The reciprocal of magnitude-response at the LCL filter resonant frequency corresponds to the grid-connected system magnitude margins (GM). Thus, Equation (25) can be written as:

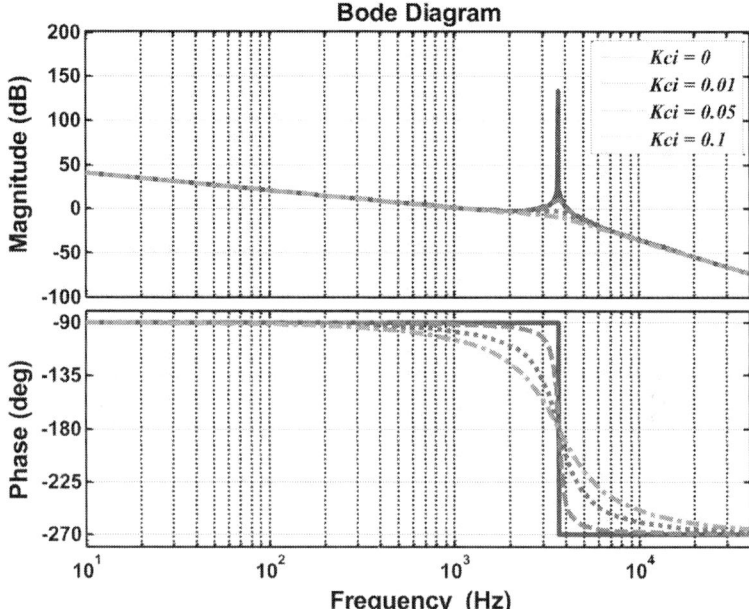

Figure 9. System with no PR controller: Bode diagram.

$$-20\lg|G(j2\pi f_r)| = -20\lg\left|\frac{K_{gi}K_{PWM}G_{PR}(j2\pi f_r)}{(j2\pi f_r)^3 L_1 L_2 C + (j2\pi f_r)^2 L_2 C K_{ci} K_{PWM} + j2\pi f_r(L_1 + L_2)}\right| \quad (25)$$
$$\geq GM$$

Since the resonant frequency is far from the fundamental frequency, the PR controller can be considered as a proportional component. Taking Equation (16) into Equation (25), the lower threshold of K_{ci} is:

$$K_{ci} \geq \frac{2\pi f_c L_1}{K_{PWM}} 10^{\frac{GM}{20}} \quad (26)$$

The phase margins (PM) of the whole system can be written as:

$$\pi + \varphi(G(j2\pi f_c)) \geq PM$$

Thus:

$$\frac{\pi}{2} + \arctan\left[(\frac{K_r}{K_p} + 1)\frac{2\omega_{PRc}\omega_c}{\omega_1^2 - \omega_c^2}\right] - \arctan(\frac{2\omega_{PRc}\omega_c}{\omega_1^2 - \omega_c^2}) - \arctan\left[\frac{K_{ci}K_{PWM}\omega_c}{L_1(\omega_r^2 - \omega_c^2)}\right] \geq PM \quad (27)$$

The superior threshold of K_{ci} can be written as Equation (28):

$$K_{ci} \leq \frac{L_1(\omega_r^2 - \omega_c^2)\tan(\frac{\pi}{2} + \arctan[(\frac{K_r}{K_p} + 1)\frac{2\omega_{PR}c\omega_c}{\omega_1^2 - \omega_c^2}] - \arctan(\frac{2\omega_{PR}c\omega_c}{\omega_1^2 - \omega_c^2}) - PM)}{K_{PWM}\omega_c} \tag{28}$$

According to Equations (26) and (28) the range of the K_{ci} is given by Equation (29):

$$\frac{\omega_c L_1}{K_{PWM}} 10^{\frac{GM}{20}} \leq K_{ci}$$

$$\leq \frac{L_1(\omega_r^2 - \omega_c^2)\tan(\frac{\pi}{2} + \arctan[(\frac{K_r}{K_p} + 1)\frac{2\omega_{PR}c\omega_c}{\omega_1^2 - \omega_c^2}] - \arctan(\frac{2\omega_{PR}c\omega_c}{\omega_1^2 - \omega_c^2}) - PM)}{K_{PWM}\omega_c} \tag{29}$$

According to Equations (24) and (29) the relationship between f_c, K_r and K_{ci} can be represented in a 3-D plot. Figure 10 shows the relationship between the three above-mentioned parameters; the space surrounded by Surface A and Surface B indicates the satisfactory range of the K_r and K_{ci} at a specific f_c:

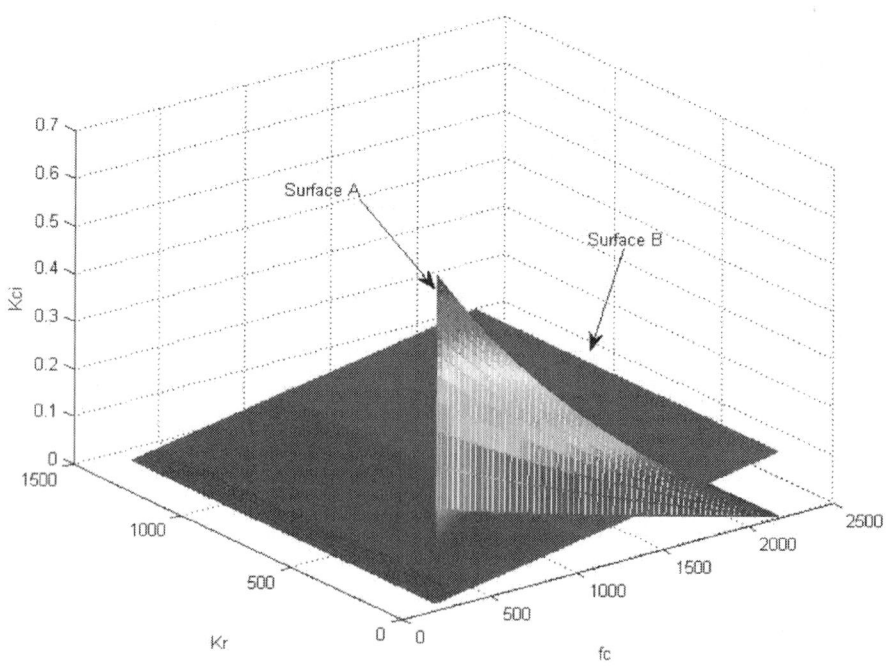

Figure 10. Relationship between f_c, K_r and K_{ci}.

NUMERICAL SIMULATION AND EXPERIMENTAL RESULTS

Following Table 1 summarizes simulation and experimental parameters.

Table 1. System parameters.

Name	Value
Fundamental frequency f_1	50 Hz
Switching frequency f_s	12 kHz
Grid phase voltage (RMS)	220 V
Grid current factor feedback K_{gi}	0.04
LCL filter inductor L_1	1.88 mH
LCL filter inductor L_2	0.34 mH
LCL filter capacitor C	6.6 µF

This section describes a practical example using parameters of Table 1. According to parameters, the calculated resonant frequency of LCL filter is 3.65 kHz. The cutoff frequency is defined as fc<fs10, and the selected value of f_c is 900 Hz.

(1). The calculated proportional gain K_p (Equation (16)) is:

$$K_p \approx \frac{2\pi f_c(L_1 + L_2)}{K_{gi}K_{PWM}} = 0.83$$

(2). The resonant factor K_r range has been calculated (Equation (24)): Supposing the steady-state tracking error of i_g is 1%, the minimum of the magnitude-frequency response of open-loop transfer function at the fundamental frequency f_1 equals 99. Assuming the minimum phase-frequency response of PR controller at the cutting frequency $-\phi$ is $-10°$. Therefore, resonance factor K_r range is:

$$Kr\ min < Kr < Kr\ max$$

where $K_{rmin} = 3.717$, and $K_{r\,max} = 41.075$.

(3) The active-damping factor K_{ci} range has been calculated through Equation (29):

Supposing the magnitude margin of the system equals 3 dB, and the phase margin 45°, thus:

$$Kci\ min < Kci < Kci\ max$$

where $K_{ci\ min} = 0.039$, $K_{ci\ max} = 0.364$.

Considering the previous three steps, a set of proper parameters can be defined:

$$Kp=0.8, Kr=20, Kci=0.04$$

Moreover, the calculated feedforward factor of the grid voltage K_{ug} equals 0.44.

MATLAB/Simulink toolbox helps verifying the feasibility of design parameters method. Figure 11shows the system Bode diagram, based on the above-mentioned parameters. The magnitude margin (3.31 dB) and the phase margin (79.6 degrees), both confirm the stability of the system.

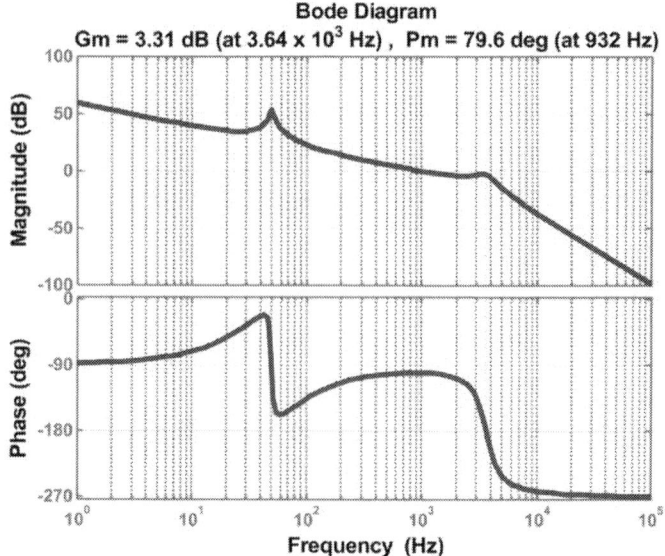

Figure 11. System Bode diagram.

However, it is impossible to manufacture a filter whose parameters completely match the initial design. Moreover, parameters change while filter operates; the grid impedance must be considered, since inductive impedance of long cables and low power transformers. Therefore, the current controller parameters must be enough robust against the parameters variation. Since the inductor L_2 is connected in series to the grid, the grid impedance variation can be merged with L_2 variation. Figures 12, 13–14 depict the Bode plots of the system. Current controller has been designed using the aforementioned method under different parameter variations.

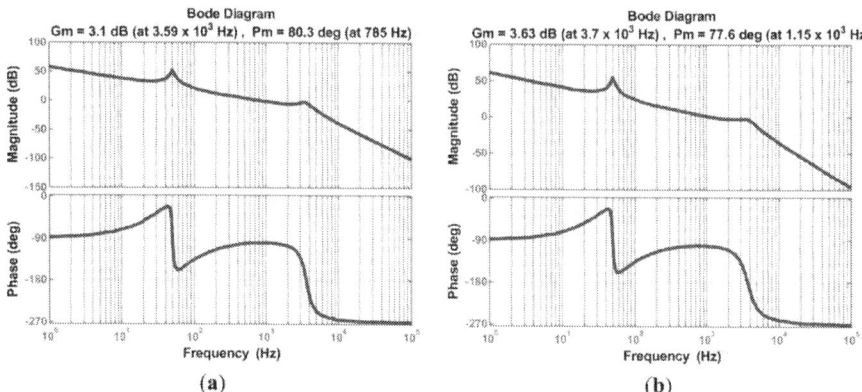

Figure 12. System Bode diagram under different parameters: **(a)** inductor L_1 increased by 20%; **(b)** inductor L_1 decreased by 20%.

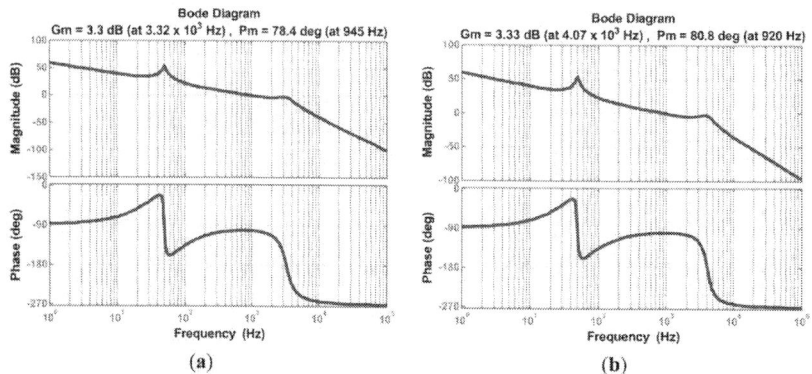

Figure 13. System Bode diagram under different parameters: **(a)** capacitor C increased by 20%; **(b)** capacitor C decreased by 20%.

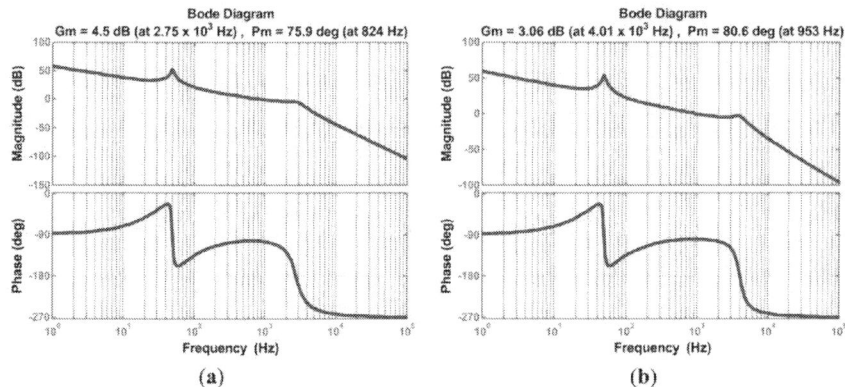

Figure 14. System Bode diagram under different parameters: **(a)** inductor L_2 increased by 150%; **(b)** inductor L_2 decreased by 20%.

Table 2 summarizes the results shown in previous Figures 12, 13–14. Since the controller design considered well the system stability margin requirements, the system is stable even with important changes in parameters, in the range −20% to 20% (in case of inductor L_2 the range extends from −20% to 250%). The minimum system magnitude margin is higher than 3 dB and the minimum phase margin is higher than 75 degrees. Therefore, the current controller parameters obtained by applying the proposed method enhance robustness against system parameters variation.

Table 2. System Magnitude and Phase margins with different parameters.

Stable margins	Nominal parameters	L_1		C		L_2	
		Increased by 20%	Decreased by 20%	Increased by 20%	Decreased by 20%	Increased by 150%	Decreased by 20%
Magnitude margin	3.31 dB	3.1 dB	3.63 dB	3.3 dB	3.33 dB	4.5 dB	3.06 dB
Phase margin	79.6 degree	80.3 degree	77.6 degree	78.4 degree	80.8 degree	75.9 degree	80.6 degree

Numerical Simulation Results

Besides the filter parameters and grid impedance variation, the grid voltage is not always stable. Therefore, we must test the robustness of the grid-connected system against grid voltage fluctuations. The application of the proposed method gives the current controller parameters. Figures 15, 16, 17–18 below show the simulation results of grid voltage fluctuation and injected current at full load under different parameter variation conditions. All simulations give the grid voltage sag equals about 42 V and voltage swell equals about 30 V. Figure 15 shows that with the nominal LCL-filter the injected current remains stable during the grid voltage sag or voltage swell. The grid fluctuation does not affect the system stability. According to the Bode diagrams shown in Figures 12, 13–14 the stable margins of the system are enough even with important changes in parameters. Figures 16, 17–18 show that the injected current is stable under nominal grid voltage with filter parameters variation. Moreover, even when the grid fluctuates, the system remains stable, as a confirmation of the effectiveness of the controller against filter parameters variation and grid fluctuations. If the inverter is connected to a weak grid (a very common situation in rural areas), the grid impedance is inductive and it can be regarded as equivalent to the increasing of inductor L_2. In this study, L_2 increases by 150% its value to simulate a weak grid situation. Even if the grid is weak and fluctuates, the overall system remains stable as Figure 17a,b shows.

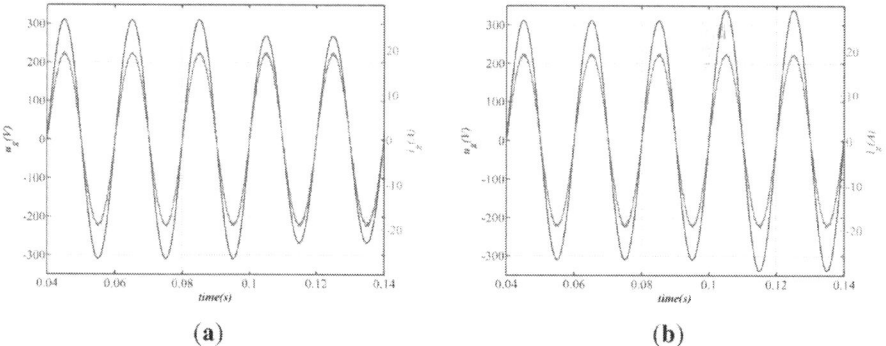

(a) (b)

Figure 15. Grid voltage and injected current at full load with nominal parameters: simulation results. (**a**) Grid voltage sag; (**b**) grid voltage swell.

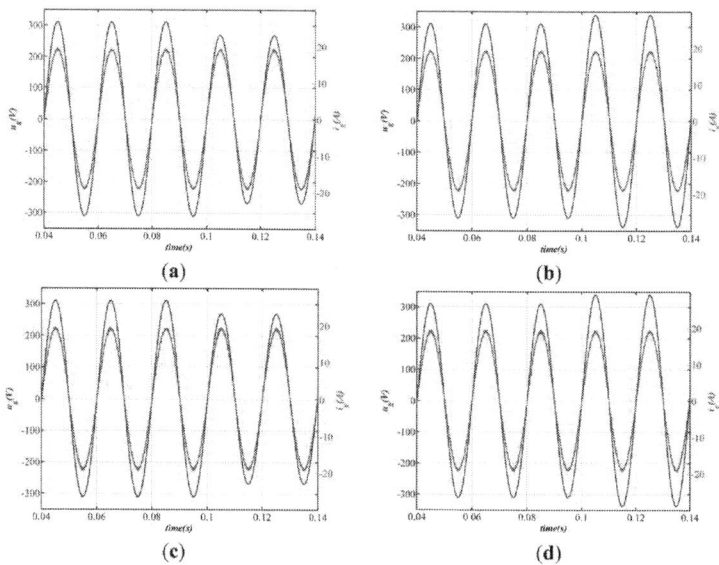

Figure   16. Grid   voltage   and   injected   current   at   full   load   with inductor L_1 variation: simulation results. (**a**) Inductor L_1increased by 20%: grid voltage sag; (**b**) Inductor L_1 increased by 20%: grid voltage swell; (**c**) Inductor L_1 decreased by 20%: grid voltage sag; (**b**) Inductor L_1 decreased by 20%: grid voltage swell.

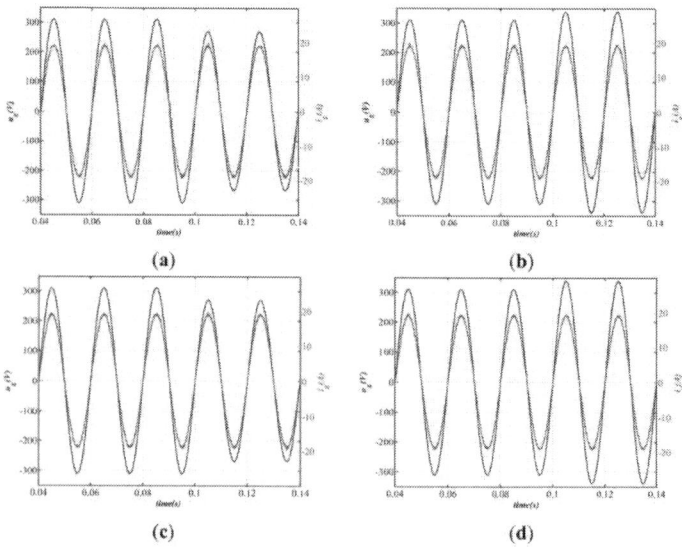

Figure   17. Grid   voltage   and   injected   current   at   full   load   with inductor L_2 variation: simulation results. (**a**) Inductor L_2increased by 150%: grid voltage sag; (**b**) inductor L_2 increased by 150%: grid voltage swell; (**c**) inductor L_2 decreased by 20%: grid voltage sag; (**b**) inductor L_2 decreased by 20%: grid voltage swell.

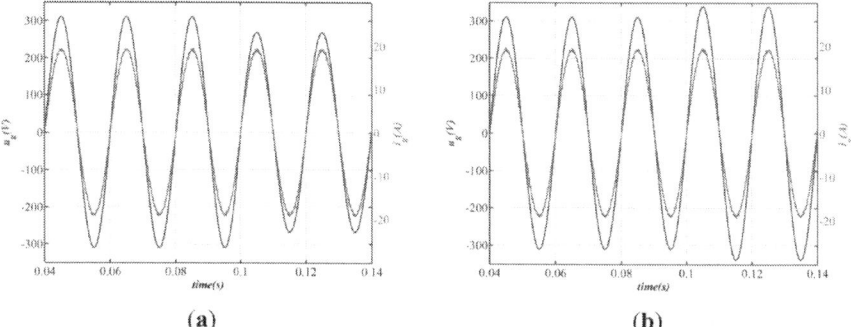

(a)   (b)

Figure   18. Grid   voltage   and   injected   current   at   full   load   with capacitor C variation: simulation results. (**a**) Capacitor C increased by 20%: grid voltage sag; (**b**) capacitor C increased by 20%: grid voltage swell; (**c**) capacitor C decreased by 20%: grid voltage sag; (**b**) capacitor C decreased by 20%: grid voltage swell.

Experimental Results

The design parameters have been verified through a 3-kW experimental setup. Inductors, capacitors, and other parameters have been set as in the corresponding simulations described in previous sections. All of the PV system algorithms have been implemented on a digital signal processor TMS320F2808 (Texas Instruments, Dallas, TX, USA). The PR controller has been discretized using the Tustin method with frequency pre-warping [17] for poles and zeroes, that yields a better matching frequency response. The adoption of the unipolar modulation SPWM regulated the injected grid current and a PV simulator replaced the PV panel. The DC source of DC-AC inverter has been given through the Boost DC-DC converter by applying the P&O MPPT method that guarantees simplicity and effectiveness [9–12]. To simulate the grid voltage fluctuation, a Programmable AC Power Source replaced the grid, while the voltage sag and swell values have been set equals to numerical simulation. Figures 19, 20, 21–22 show the experimental waveforms at full load under different parameters variations; the simulation results and the experimental data have been plot into one-to-one correspondence. Although the experimental results are worse than the numerical simulation ones because of the parasitic parameters of the inductors and capacitor, they are still very satisfactory. The experimental data well match numerical simulations.

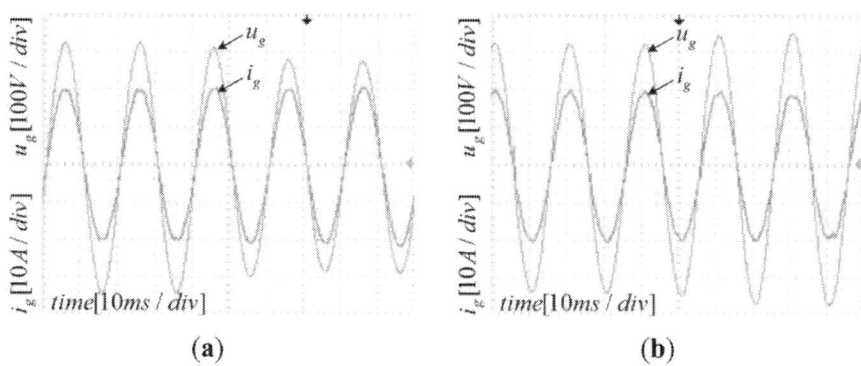

Figure 19. Grid voltage and injected current at full load with nominal parameters: experimental results. (**a**) Grid voltage sag; (**b**) grid voltage swell.

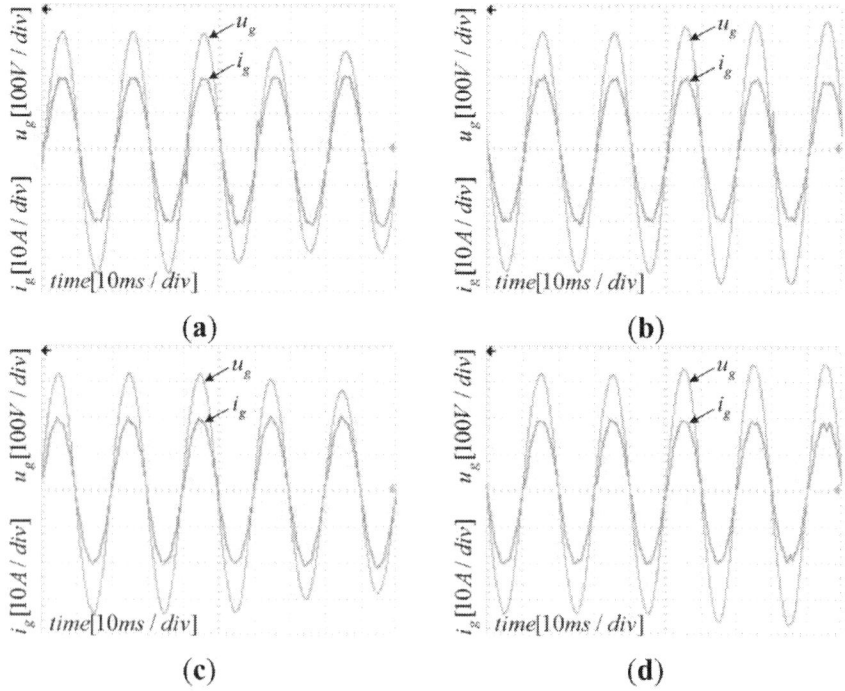

Figure 20. Grid voltage and injected current at full load with inductor L_1 variation: experimental results. (**a**) Inductor L_1 increased by 20%: grid voltage sag; (**b**) inductor L_1 increased by 20%: grid voltage swell; (**c**) inductor L_1 decreased by 20%: grid voltage sag; (**b**) inductor L_1 decreased by 20%: grid voltage swell.

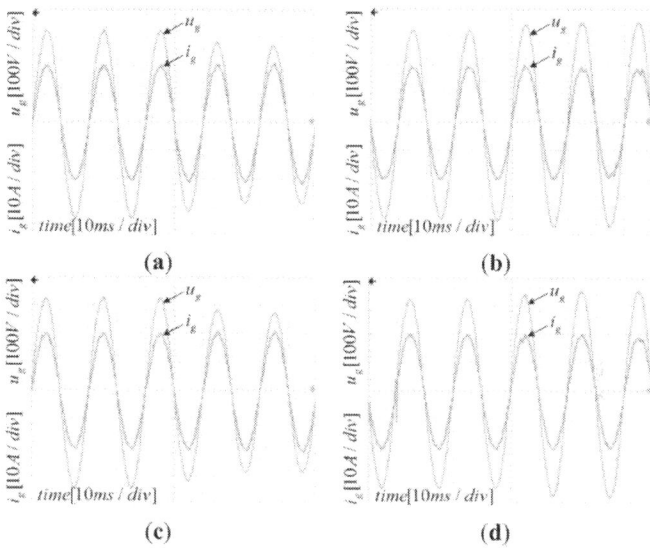

Figure 21. Grid voltage and injected current at full load with inductor L_2 variation: experimental results. (**a**) Inductor L_2 increased by 150%: grid voltage sag; (**b**) inductor L_2 increased by 150%: grid voltage swell; (**c**) inductor L_2 decreased by 20%: grid voltage sag; (**b**) inductor L_2 decreased by 20%: grid voltage swell.

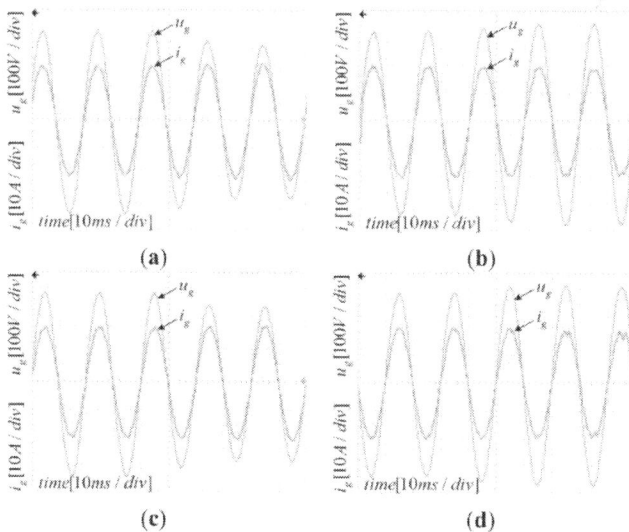

Figure 22. Grid voltage and injected current at full load with capacitor C variation: experimental results. (**a**) Capacitor C increased by 20%: grid voltage sag; (**b**) capacitor C increased by 20%: grid voltage swell; (**c**) capacitor C decreased by 20%: grid voltage sag; (**b**) capacitor C decreased by 20%: grid voltage swell.

CONCLUSIONS

The stability analysis of the system composed by a PR controller and an LCL filter together is not easy: the frequency responses may affect each other and the PR controller design becomes complex. The traditional method based on trial-and-error procedures, is too time-consuming, and the design process is inefficient. This paper provides a detailed analysis of the frequency response influence between the PR controller and the LCL filter. In addition, the paper presents a systematic design method for the PR controller parameters and the capacitor current feedback coefficient, used in the active damping of the LCL filter. Using the new parameters, a numerical simulation shows that the system meets the requirements of stable margins and current tracking steady-state error. The robustness of the current controller is verified through several experimental tests carried out on a 3 kW platform varying the system parameters. The Bode diagrams of the system varying inductor, capacitor, and grid impedance values confirmed that the controller parameters enhance robustness against the system parameters variation. Moreover, the system remains stable even in case of grid voltage fluctuation. Both the simulation and the experimental results assess the validity of the proposed design method.

REFERENCES

1. Carrasco, J.M.; Franquelo, L.G.; Bialasiewicz, J.T.; Galvan, E.; Guisado, R.C.P.; Prats, A.M.; Leon, J.I.; Moreno-Alfonso, N. Power-electronic systems for the grid integration of renewable energy sources: A survey. *IEEE Trans. Ind. Electron.* 2006, *53*, 1002–1016.
2. Wessels, C.; Dannehl, J.; Fuchs, F.W. Active Damping of LCL-Filter Resonance based on Virtual Resistor for PWM Rectifiers—Stability Analysis with Different Filter Parameters. Proceedings of the 2008 IEEE Power Electronics Specialists Conference, Rhodes, Greece, 15–19 June 2008; pp. 3532–3538.
3. Castilla, M.; Miret, J.; Matas, J.; de Vicuna, L.G.; Guerrero, J.M. Control design guidelines for single-phase grid-connected photovoltaic inverters with damped resonant harmonic compensators. *IEEE Trans. Ind. Electron.* 2009, *56*, 4492–4501.

4. Yi, L.; Zhengming, Z.; Fanbo, H.; Sizhao, L.; Lu, Y. An Improved Virtual Resistance Damping Method for Grid-Connected Inverters with LCL Filters. Proceedings of the 2011 IEEE Energy Conversion Congress and Exposition (ECCE 2011), Phoenix, AZ, USA, 17–22 September 2011; pp. 3816–3822.

5. Parker, S.G.; McGrath, B.P.; Holmes, D.G. Regions of Active Damping Control for LCL Filters. Proceedings of the Energy Conversion Congress and Exposition (ECCE), Raleigh, NC, USA, 15–20 September 2012; pp. 53–60.

6. Bao, C.L.; Ruan, X.B.; Wang, X.H.; Li, W.W.; Pan, D.H.; Weng, K.L. Design of Injected Grid Current Regulator and Capacitor-Current-Feedback Active-Damping for LCL-Type Grid-Connected Inverter. Proceedings of the Energy Conversion Congress and Exposition (ECCE), Raleigh, NC, USA, 15–20 September 2012; pp. 579–586.

7. Fukuda, S.; Yoda, T. A novel current-tracking method for active filters based on a sinusoidal internal model. *IEEE Trans. Ind. Appl.* 2001, *37*, 888–895.

8. Zmood, D.N.; Holmes, D.G. Stationary frame current regulation of PWM inverters with zero steady-state error. *IEEE Trans. Power Electron.* 2003, *18*, 814–822.

9. Esram, T.; Chapman, P.L. Comparison of photovoltaic array maximum power point tracking techniques. *IEEE Trans. Energy Convers.* 2007, *22*, 439–449.

10. Noguchi, T.; Togashi, S.; Nakamoto, R. Short-current pulse-based maximum-power-point tracking method for multiple photovoltaic-and-converter module system. *IEEE Trans. Ind. Electron.* 2002,*49*, 217–223.

11. Mutoh, N.; Ohno, M.; Inoue, T. A method for MPPT control while searching for parameters corresponding to weather conditions for PV generation systems. *IEEE Trans. Ind. Electron.* 2006,*53*, 1055–1065.

12. Petrone, G.; Spagnuolo, G.; Vitelli, M. A multivariable perturb-and-observe maximum power point tracking technique applied to a single-stage photovoltaic inverter. *IEEE Trans. Ind. Electron.*2011, *58*, 76–84.

13. Ciobotaru, M.; Teodorescu, R.; Blaabjerg, F. A New Single-Phase PLL Structure Based on Second Order Generalized Integrator. Proceedings of the 2006 IEEE Power Electronics Specialists Conference (IEEE Cat. No. 06CH37819C), Jeju, Korea, 18–22 June 2006; pp. 1–6.

14. Silva, S.M.; Lopes, B.M.; Cardoso, B.J.; Campana, R.P.; Boaventura, W.C. Performance Evaluation of PLL Algorithms for Single-Phase Grid-Connected Systems. Conference Record of the 2004 IEEE Industry Applications Conference, Seattle, WA, USA, 3–7 October 2004; Volumes 1–4, Covering Theory to Practice. pp. 2259–2263.

15. Zmood, D.N.; Holmes, D.G.; Bode, G.H. Frequency-domain analysis of three-phase linear current regulators. *IEEE Trans. Ind. Appl.* 2001, *37*, 601–610.

16. Dannehl, J.; Wessels, C.; Fuchs, F.W. Limitations of voltage-oriented PI current control of grid-connected pwm rectifiers with LCL filters. *IEEE Trans. Ind. Electron.* 2009, *56*, 380–388.

17. Yepes, A.G.; Freijedo, F.D.; Doval-Gandoy, J.; Lopez, O.; Malvar, J.; Fernandez-Comesana, P. Effects of discretization methods on the performance of resonant controllers. *IEEE Trans. Power Electron.* 2012, *27*, 4976.

CITATION

Ningyun Zhang, Houjun Tang and Chen Yao, A Systematic Method for Designing a PR Controller and Active Damping of the LCL Filter for Single-Phase Grid-Connected PV Inverters, doi:10.3390/en7063934

CHAPTER 8

Converter Controls and Flicker Study of PMSG-Based Grid Connected Wind Turbines

Ali H. Kasem Alaboudy[1], Ahmed A. Daoud[2], Sobhy S. Desouky[2] and Ahmed A. Salem[3]

[1]Faculty of Industrial Education, Suez Canal University, Suez Campus, Suez, Egypt
[2]Faculty of Engineering, Port-Said University, Port-Said, Egypt
[3]Faculty of Engineering, Suez Canal University, Ismailia 41522, Egypt

ABSTRACT

With the increased penetration of wind power, the influence of wind turbine generators on the grid power quality stipulates careful investigation and analysis. Direct driven permanent magnet synchronous generator (PMSG) with a back-to-back converter set is one of the promising technologies in wind power generation schemes. In this paper, comprehensive models of wind turbine are used to analyze power and voltage fluctuations. The short time flicker index is used to assess the voltage fluctuation emitted. The control scheme of the grid-side converter is supported with a voltage regulation loop to reduce flicker emission. The effects of grid and site parameters on voltage fluctuation are investigated. Simulation results show that reduced flicker emissions are given when the developed voltage regulation loop is activated. Reasonable values of grid and site parameters contribute in the minimization of voltage fluctuation and flicker emission levels.

INTRODUCTION

Nowadays, wind power generators represent a prominent facility for generating renewable and clean bulk power to utility grids. Basically, there are many good reasons for using more wind energy on power grids. For instance, wind generation is supported by not only being clean and renewable but also having minimal running cost requirements [1].

The amount of the energy extracted from the wind depends not only on the incident wind speed, but also on the control system applied on the wind energy conversion system (WECS). Typically, maximum wind power extraction is accomplished by using fully controlled variable speed wind turbine generators. The rotational speed of wind turbine hub is adjusted according to the incident wind speed to track the maximum wind power trajectory [2]. As a power generating unit connected to the grid, the wind turbine generator should have the capability to control the active and reactive powers injected to grid. Variable speed wind turbine (VSWT) topologies include many different generator/converter configurations, based on cost, efficiency, annual energy capturing, and control complexity of the overall system [3].

Permanent magnet synchronous generator (PMSG) based variable speed wind turbines are considered appropriate and feasible technology in wind generation industry since PMSGs are self-excited, and thus allows operation at high power factor and high efficiency [4], [5] and [6]. Furthermore, due to its low rotational speed the gearbox can be omitted; where in other WECSs, the gearbox is one of the most critical turbine components, since its failure is highly expected, and, hence, it requires careful and regular maintenance [7].

Due to the stochastic nature of wind, power and voltage generated by a wind turbine are more variable than that produced by conventional generators. With the increased wind power penetration into the grid, the influence of wind turbines on the power quality becomes an important issue, and, hence, IEC 61400-21 is developed to provide procedures for determining the power quality characteristics of wind turbines [8].

One of the most important wind-power quality considerations is the effect of voltage fluctuation. The lighting flicker level is generally used to measure voltage fluctuation. Voltage fluctuation disturbs the sensitive electric and electronic equipment. This may lead to a great reduction in the life span of most equipment, flicker has widely been considered as a serious drawback and may limit for the maximum amount of wind power generation that can be connected to the grid [9].

Flicker is induced by voltage fluctuations, which are caused by load flow changes in the grid. The flicker emission produced by grid-connected variable-speed wind turbines with full-scale back-to-back converters during continuous operation is mainly caused by fluctuations in the output power. The output power variation results from wind speed variations, the wind shear, and the tower shadow effects [8]. The wind shear and the tower shadow effects are normally referred to as the 3p oscillations. As a consequence, an output power drop will appear three times per revolution for a three-bladed wind turbine [8], [9], [10] and [11].

There are numerous factors that affect flicker emission of grid-connected wind turbines during continuous operation, such as wind characteristics (e.g. mean wind speed, turbulence intensity) and grid conditions (e.g. short circuit capacity, grid impedance angle) [8], [9], [10] and [12]. The type of wind turbine also has an influence on flicker emission. Variable speed wind turbines have shown better performance related to flicker emission in comparison with constant speed wind turbines [10] and [13]. The better performance of VSWT is due to (1) the buffering effect of the back-to-back converter set, and (2) rotor speed flexibility that converts power spikes into speed variations. Although the variable speed wind turbine produces lower flicker levels, the flicker study becomes necessary and imperative as the wind power penetration is continuously increasing. Several articles present flicker mitigation techniques for induction generator based wind turbines [12] and [14]. Recently published articles are devoted to address the flicker emission caused by variable-speed PMSG-based wind turbines [11] and [15]. The flicker mitigation techniques proposed in these articles [11] and [15] are based on tuning active and reactive powers. In this paper, a modified grid-side converter control scheme is developed to minimize flicker emission.

This paper tackles the performance of grid connected direct driven PMSG based wind turbines under continuous operation. The PMSG is connected to the grid at the point of common coupling (PCC) via an ac–dc–ac back-to-back converter set. Two control schemes are developed for generator- and grid-side converters. The control of the generator side converter is developed to achieve maximum power point tracking (MPPT), while, the control scheme of the grid side converter is designed to operate at unity power factor and stabilize the dc link voltage to its nominal value. The control scheme of the grid-side converter is supported with a voltage regulation loop to reduce flicker emission. Based on the wind turbine model, flicker emission of variable speed PMSG-based wind turbines is investigated during continuous operation. The influence of different factors on flicker caused by wind turbines are examined on normal operation with and without voltage regulation control loop. Comprehensive models of the WECS and flickermeter considering system dynamics, and control actions are presented in MATLAB/SIMULINK environment.

SYSTEM MODELING

Wind energy conversion system converts kinetic energy of the wind to mechanical energy by means of wind turbine rotor blades then the generator converts the mechanical power to electrical power that is being fed to the grid through power electronic converters. The WECS under study, described in Fig. 1, consists of two main parts

(a). Mechanical parts: include the aerodynamic system with the rotor blades and the drive train system (if existed).
(b). Electrical parts: comprised of the PMSG and the back-to-back converter set [16].

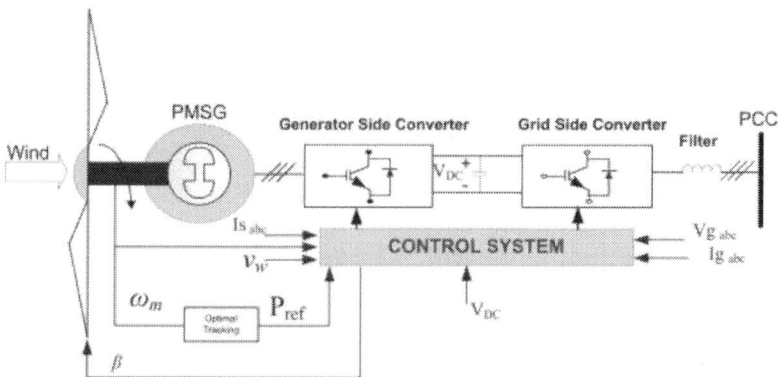

Figure 1. PMSG based wind turbine configuration.

Mechanical Part Representation

The mechanical part is responsible for converting wind power incident on turbine rotor blades which rotate with swept area ($A = \pi r^2$), to mechanical power form. The mechanical power is given as follows:

$$P_m = \frac{1}{2} C_p \rho \pi r^2 V_w^3 \tag{1}$$

where ρ is the air density (kg/m^3), r is the blade length (m), V_w is wind speed (m/s) and C_p is the power coefficient. The value of C_p is dependent on the tip speed ratio (λ) and the blades pitch angle (β). The blades pitch angle is adjusted by the embedded pitch controllers and depends on the type and operating condition of the wind turbine. The mathematical expression of C_p is given in (2) [17].

$$C_p = c_1 \left(\frac{c_2}{\lambda_i} - c_3 \beta - c_4 \right) e^{\frac{-c_5}{\lambda_i}} + c_6 \lambda \tag{2}$$

The parameters λ_i and c_1–c_6 are defined in the Appendix [17].
The tip speed ratio is given as follows:

$$\lambda = \frac{\omega_t r}{V_w} \tag{3}$$

where ω_t (rad/s) is the rotational speed of the turbine shaft.
The mechanical input torque, T_m, is given as follows:

$$T_m = \frac{P_m}{\omega_t} \tag{4}$$

The operating region of wind turbines can be divided into three states – standstill region, normal operation region and pitched operation region, as seen in Fig. 2.

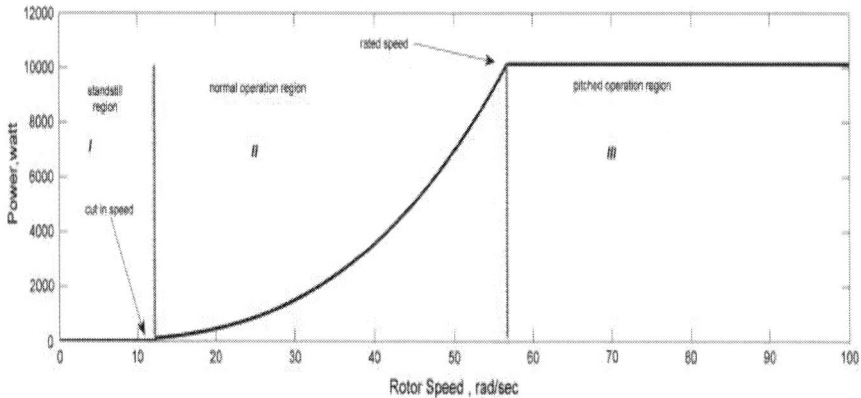

Figure 2. Turbine operating region.

In standstill region, the wind turbine rotor is blocked since the wind power is not worth and could be insufficient to overcome the friction and inertia of the wind turbine. At the cut-in speed (indicated at the end border of standstill region), the turbine starts to run. In this work, the cut-in wind speed is taken to be 3 m/s, which is equivalent to $\omega_t = 12.15$ rad/s.

However, at normal operation region, the turbine starts at cut in wind speed to follows the MPPT approach, ideally, the wind turbine should be operated at maximum C_p most of the time with fixed pitch angle, $\beta = 0°$. As illustrated in Fig. 3, the C_p–λ family of curves for different pitch angles is given, where the maximum value of the power coefficient, $C_{p\_max} \approx 0.48$, is given at $\beta = 0°$ and $\lambda = \lambda_{opt} \approx 8.1$. This region ends when the wind speed reaches its rated value of 14 m/s that is equivalent to $\omega_t = 56.7$ rad/s, as seen in Fig. 4 that describes the wind turbine output power characteristics. Based on Fig. 4, for any particular wind speed there is optimum rotational speed, which gives the maximum power capturing.

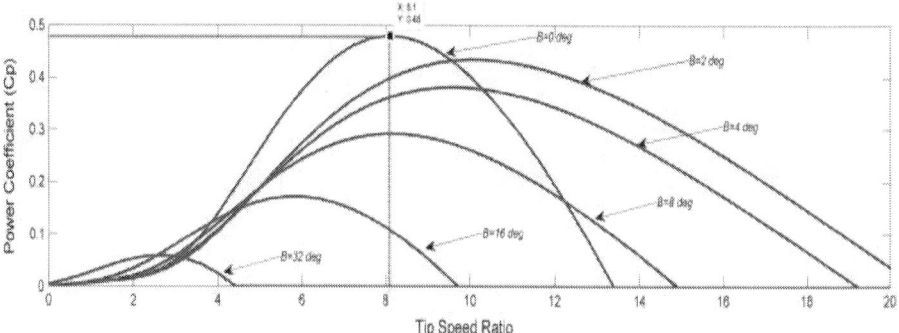

Figure 3. C_p–λ curves for different values of the pitch angle, β.

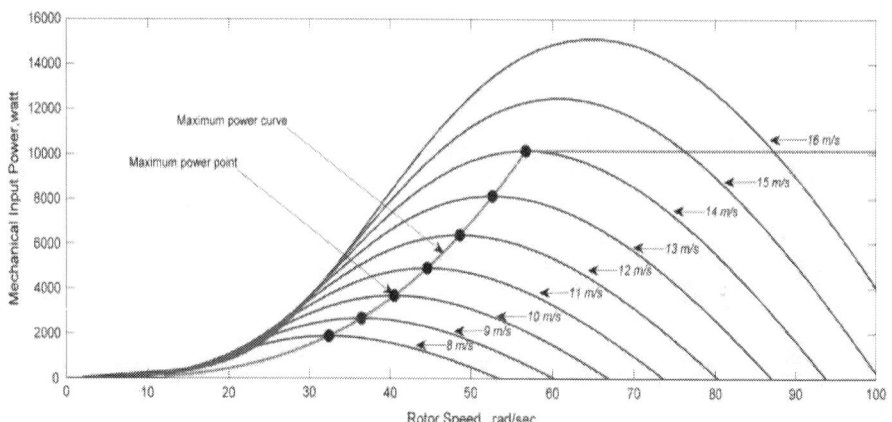

Figure 4. The power characteristic of the wind turbine used in this study.

Finally, at pitched operation region, the power captured by blades is limited by the action of pitch control mechanism. As the wind speed increases, the power generated by the wind turbine also increases. Once the turbine power rating is reached, (indicated by the rated wind speed), the pitch angle must be increased to shed the aerodynamic power. As the pitch angle is increased, and hence, the wind turbine operates at lower efficiency.

Fig. 5 shows how the pitch angle changes when wind speed increases. From cut in wind speed (3 m/s) to rated wind speed (14 m/s), the blade is kept at fixed angle $\beta = 0°$ to achieve MPPT, However, when the wind speed exceeds 14 m/s, the pitch control rotate all or part of each blade about its axis with particular angle to limit the power to its rated value. The standard wind speed model, presented in [18], is employed to emerge the wind speed variation at certain mean wind speed and turbulence values.

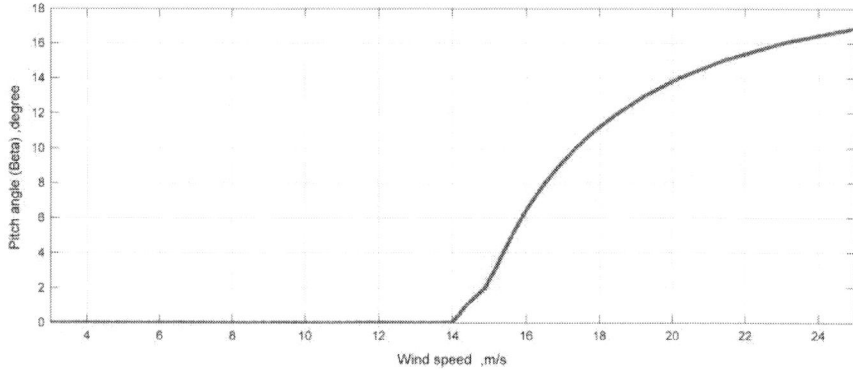

Figure 5. Pitch angle for different wind speed values.

Electrical Part Representation
PMSG Modeling
PMSG converts the mechanical power from aerodynamic system to ac electrical power, which is then converted to dc power through IGBT pulse width modulation (PWM) converter connected with dc link at its dc port. The power is transferred to the grid through another IGBT pulse width modulation (PWM) inverter.

The electrical model of the surface mounted PMSG has been developed in [2] and [4]. It is typically implemented in the *dq* rotating reference frame. The equivalent circuits of the PMSG in direct and quadrature axes are shown in Fig. 6. The stator voltage equations in the *d-q* reference frame, V_{sd} and V_{sq}, are given as follows.

$$V_{sd} = -R_s I_{sd} - L_s \frac{d}{dt} I_{sd} + L_s \omega_e I_{sq} \tag{5}$$

$$V_{sq} = -R_s I_{sq} - L_s \frac{d}{dt} I_{sq} - L_s \omega_e I_{sd} + \omega_e \phi \tag{6}$$

where L_S, and R_S respectively represent the inductance and resistance of the PMSG winding, ϕ represents the magnet flux, ω_e is the electrical rotational speed of generator, and I_{sd}, I_{sq} are the direct and quadrature components of the machine currents respectively.

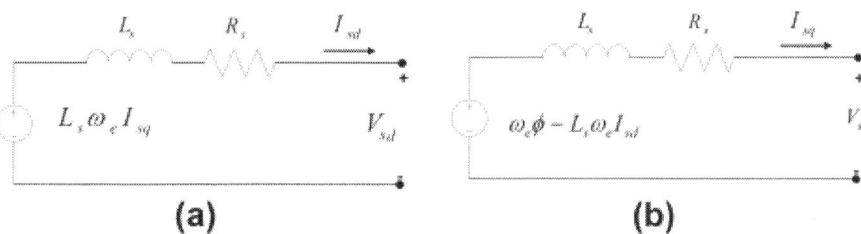

Figure 6. Equivalent circuits of PMSG: (a) d-axis (b) q-axis.

The electromagnetic torque, T_e, is given in [2] and [4].

$$T_e = \frac{3}{2}pI_{sq}((L_d - L_q)I_{sd} + \phi) \tag{7}$$

where L_d, L_q are the two axes machine inductances; p is the no. of pole pairs.

In surface mounted PMSGs, $L_d = L_q = L_s$. Hence, the electromagnetic torque can be rewritten as follows.

$$T_e = \frac{3}{2}p \cdot I_{sq} \cdot \phi \tag{8}$$

The relation between angular frequency of the stator voltage (electrical angular velocity)ω_e, and mechanical angular velocity of the generator rotor ω_m is given below [5]:

$$\omega_e = \omega_m p \tag{9}$$

Interfacing Converters
As shown in Fig. 1, in variable speed wind turbines, PMSG is connected to the utility grid via a back-to-back set of converters. The first converter, known as the generator side converter, is connected to the stator windings of the PMSG. While the other one is known as grid side converter and is

connected to the grid at the PCC via ac filter. The dc terminals of the two converters are collected together with shunt dc capacitor. The power scheme of the each converter simply contains a three-leg voltage source inverter. However, from control prospective, different control schemes based on the control functions can be applied on the inverter switches. Control schemes of the two converters will be described in details in the following sections.

CONTROL OF THE GENERATOR SIDE CONVERTER

The generator side converter control is mainly used to control the wind turbine shaft speed in order to maximize the output power. In a variable speed wind energy conversion system, the maximum power at different wind speeds depends on the power coefficient, C_p. Unfortunately, for wind turbines, C_p is not constant. The parameters affecting the coefficient C_p are: the tip speed ratio λ and the pitch angle β as illustrated in Fig. 3. To obtain the maximum power production (P_{max}) from the wind turbine, the turbine should operate at C_{p_max} and hence, it is necessary to keep the generator rotor speed ω_m to meet the optimum value of the tip speed ratio (λ_{opt}). If the wind speed varies, the rotor speed should be adjusted to follow the change of the wind speed [19], according to the maximum power curve shown in Fig. 4. The generator speed control is typically accomplished through the generator side converter. Hence, the control of the generator-side converter allows the generator to tune the rotational speed depending on the incident wind variation.

To understand the control concept, the equation of motion should be discussed. The motion equation for a typical generator is given as follows [6].

$$J\frac{d\omega_m}{dt} = T_m - T_e - B\omega_m$$

(10)

where J is the inertia of the whole system including the turbine and generator (Kg m^2), while B friction factor (N m s).

The mechanical rotational speed of PMSG rotor is given by:

$$\omega_m = \omega_t G_r$$

(11)

where ω_t is the turbine rotational speed and G_r gear ratio (if existed). For gearless PMSG based wind turbine $G_r = 1$.

Based on (10), the generator rotational speed is governed by the electromagnetic torque, and hence speed control is obtained by generator torque control. From (8), the electromagnetic torque may be controlled directly by q-axis current component, I_{sq}, and hence, the speed can be controlled by changing the q-axis current component. The d-axis current component I_{sd} is set to zero to minimize the current flow for a given torque, and thus minimizes the resistive losses [4].

The stator voltage components, V_{sd}, and V_{sq}, synthesized by the generator side converter, can be employed to govern the generator current components, I_{sd}, and I_{sq}, as seen in Eqs. (5) and (6). The controller requires feedback from the PMSG stator current components, I_{sq}, I_{sd}. The error between measured and reference components is the input of PI controller [5]. Then, compensation terms shown in Eqs. (5) and (6) are considered to ensure stable and decoupled active and reactive power control as shown in Fig. 7.Then the output voltage will be the input of space vector modulation (SVM) to produce switching signal to drive the generator side converter.

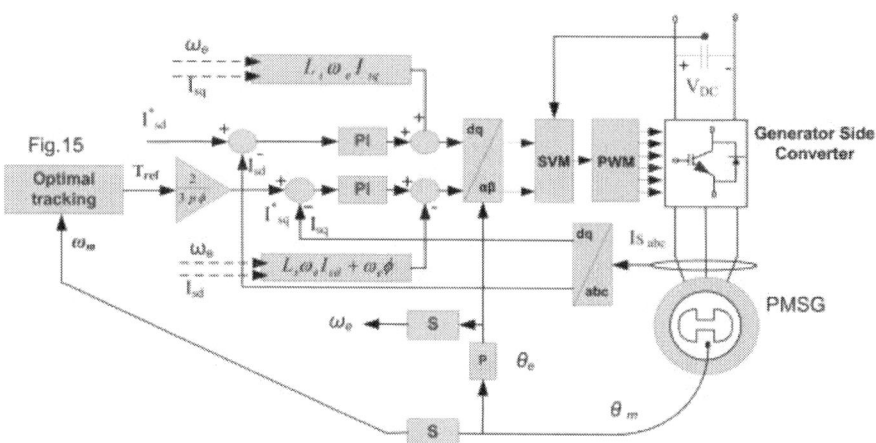

Figure 7. Block diagram of generator side converter.

CONTROL OF THE GRID SIDE CONVERTER

The objective of the grid side converter control is to stabilize the dc link voltage at its nominal value 750 V. Choosing the DC link voltage (V_{DC}) is related to the value of the LL rms voltage seen in the grid side ($V_{g\text{-}LL}$). The relation is: $V_{DC} \geqslant 1.633 V_{g\text{-}LL}$, as declared[20]. Consequently, ensuring that the active power generated by the generator is fed to the grid where the capacitor voltage always varies during wind turbine operation [21]. It's worth mentioned that there are many control strategies used to perform grid side converter depending on the reference frame used to perform control strategies. Here, the synchronous reference frame control strategy is under investigation. The dynamic model of the grid connection, in reference frame rotating synchronously with the grid voltage, is given as follows [4]:

$$V_{gd} = V_{id} - RI_{gd} - L\frac{d}{dt}I_{gd} + L\omega_g I_{gq} \quad (12)$$

$$V_{gq} = V_{iq} - RI_{gq} - L\frac{d}{dt}I_{gq} - L\omega_g I_{gd} \quad (13)$$

where L and R are the grid inductance and resistance, respectively. V_{id} and V_{iq} are the inverter voltage components. If the reference frame is oriented along the supply voltage, the grid vector voltage is:

$$V = V_{gd} + j0 \quad (14)$$

Active and reactive power can be expressed as follows [4]. equation(15)

$$P_g = \frac{3}{2}V_{gd}I_{gd} \quad (15)$$

$$Q_g = \frac{3}{2}V_{gd}I_{gq} \quad (16)$$

It could be seen from above equations that we can control the active and reactive powers by respectively changing the d and q-current components. Also in order to transfer all the active power generated by the wind turbine the dc-link voltage must remain constant [21], as explained in the following constraint [22].

$$C\frac{dV_{DC}}{dt} = \frac{P_t}{V_{DC}} - \frac{P_g}{V_{DC}} \tag{17}$$

where subscript 'g' refers to the grid and 't' refers to the wind turbine.

Based on (17), if the two powers (the wind turbine power and the grid power) are equal there will be no change in the dc-link voltage.

The control strategy of the grid side converter (given in Fig. 8) contains two cascaded loops. The inner loops control the grid currents and the outer loops control the dc-link voltage and the reactive power. The current control loops could be employed to dismiss the power quality issues, thus harmonic compensation are inserted to the action of the current controllers to improve it. The outer loops regulate the power flow of the system by controlling the active and reactive powers delivered to the grid. Further, unity power factor flow (zero reactive power exchange) could be easily obtained, unless the grid operators require different reactive power settings.

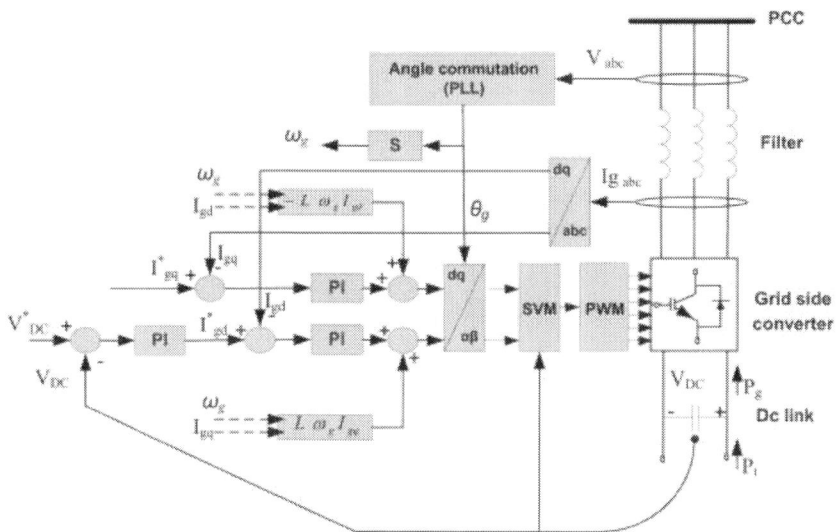

Figure 8. Block diagram of grid side converter.

In this control strategy, the currents are represented in the dq synchronous rotating reference frame and controlled with standard PI controllers. This control transforms the grid voltages and currents from the abc to their equivalents in the dq reference frame. Worth mentioning that the abc variables are transformed to dc components (dq components), and thus offers much easier and more feasible controllability. The control structures developed in this work, use PI controllers since they have proper performance for controlling dc variables [22]. It could be seen from Fig. 8 that the outer loops control the dc voltage by taking the dc voltage reference of 750 V, and the error signal produce I_{gd} reference to inner current control loop that control active power. The second channel controls the reactive power by producing I_{gq} reference to inner current control loop. The reactive power is set equal to zero.

SIMULATION RESULTS

To examine the performance of the implemented control scheme, two case studies considering different wind speed variations have been conducted; the parameter of the system is defined in the appendix:

Case (1)

It is assumed that the wind speed profile varies up and down as step function with mean wind speed 11.4 m/s, as seen from Fig. 9a. The time span is 5 s. The controller adjusts the PMSG rotational speed as the wind speed varies to maintain the optimal tip speed ratio, λ_{opt}, as depicted in Fig. 9b. The variation of C_p with time is demonstrated in Fig. 9c. Irrespective of wind speed variation, the power coefficient is maintained to its maximum value $C_{p\ max}$. At wind speed transitions, shown $t = 0.75$ s, 2 s, and 3.5 s there are some minor notches in the C_p variation, due to the system inertia.

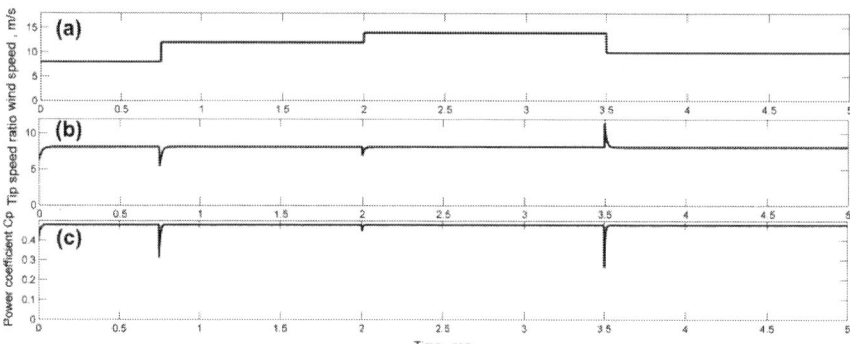

Figure 9. (a) Wind speed variation in m/s, (b) Tip speed ratio λ, and (c) power coefficient C_p.

As shown in Fig. 10a, there is a good agreement between the actual and reference values of the turbine rotor speed. The generator is initially rotating at 25 rad/s. As shown in Fig. 10b, when the wind speed increases, the input mechanical power also increases and thus the electrical power produced by the generator increases. Typically, the mechanical input power is slightly greater than the electrical power production due to the system losses.

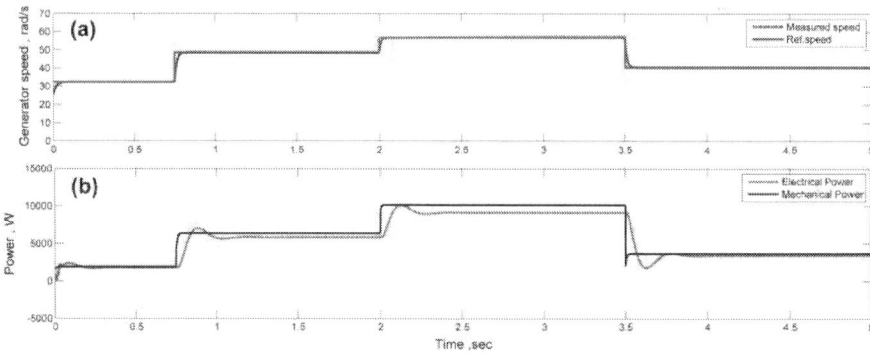

Figure 10. (a) Measured and reference speed of PMSG in rad/s, and (b) Mechanical input and output Electric Power in Watt.

To realize the feasibility of the grid side controller, Fig. 11 presents the dc link voltage variation, active, and reactive powers. The controller gives good agreement between the actual and reference values of the dc link voltage thus the actual dc voltage is almost constant over the whole investigation window as shown in Fig. 11a. Fig. 11b shows that the power injected to the grid is dependent on the average wind speed, The reactive power fed to the grid is shown in Fig. 11c which is approximately zero, i.e., the generator is working at unity power factor.

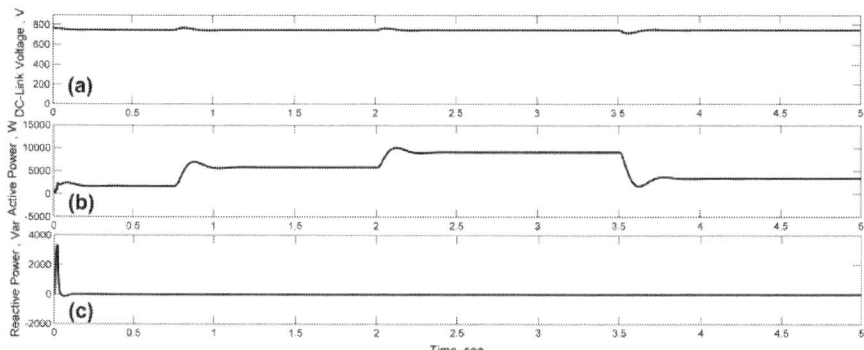

Figure 11. Simulation results of WECS under grid side converter, case (1): (a) dc link voltage in volt. (b) Grid output active power in Watts, and (c) Grid output reactive power in Vars.

Case (2)

In case (2), the wind speed profile, shown in Fig. 12a, varies randomly with time and having an average value of 12 m/s and 20% turbulence intensity. Wind speed time series is given according to the equivalent wind speed model provided in Wind Turbine Blockset, Matlab/Simulink [18]. The tip speed ratio and power coefficient of wind turbine are given in Fig. 12b and c respectively. It is obvious from these figures that λ_{opt} and C_{p_max} are almost constant for the period conducted in simulation.

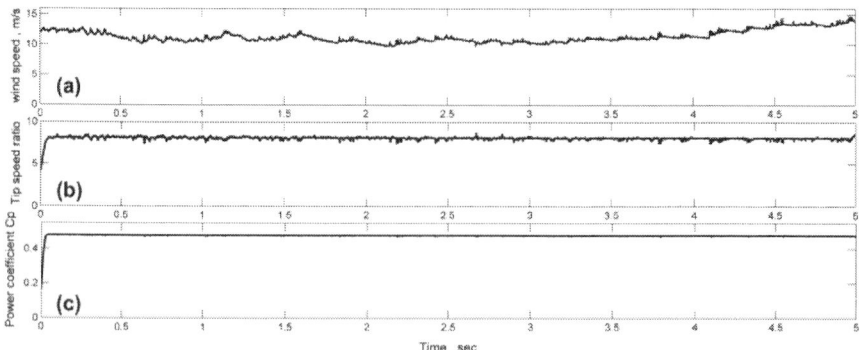

Figure 12. (a) Wind speed variation in m/s, (b) Tip speed ratio λ, and (c) Power coefficient C_p.

The generator speed and the electrical and mechanical powers are presented in Fig. 13. As shown from Fig. 13a, the actual speed almost coincides with the reference speed. Hence, the control scheme of the generator side converter gives appropriate performance and is capable to afford the MPPT facility for PMSG-based wind turbines.Fig. 13b shows the mechanical and electrical powers produced by the wind turbine. It is clear that impacts of the capricious variation in wind speed profile are more significant on the mechanical input power compared to electrical output power. This is due to the generator inertia and buffering effect of the converters with the dc-link.

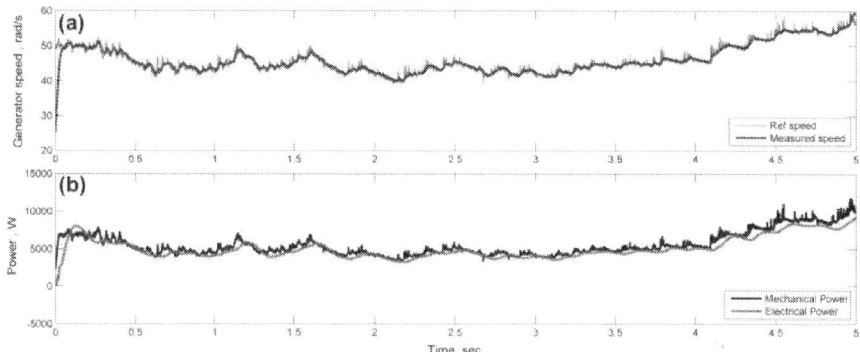

Figure 13. (a) Measured and reference speed of PMSG in rad/s, and (b) mechanical input and output electric power in Watt.

Fig. 14 gives the performance of grid side converter controller under random change in wind speed profile, as shown in Fig. 14a the dc link voltage is constant over the whole period. The active power fed to grid is shown in Fig. 14b, and the reactive power fed to grid is shown in Fig. 14c. Zero reactive power exchange at the PCC is recognized.

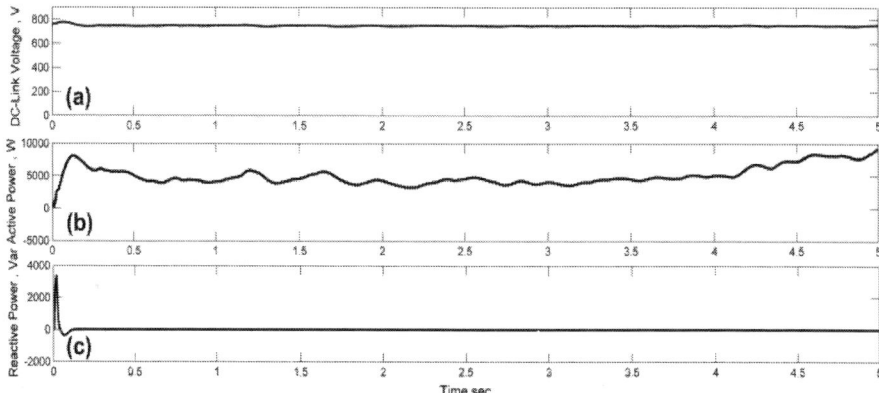

Figure 14. Simulation results of WECS under grid side converter, case (2): (a) dc link voltage in volt. (b) Grid output active power in Watts, and (c) Grid output reactive power in Vars.

IDENTIFICATION OF MPPT

MPPT is implemented without wind speed sensor. The optimality is achieved when the generator torque follows the optimum torque curve, therefore the generator speed is governed by means of generator torque control [21] and [23].

To explain the concept of MPPT for the case under study, the wind speed profile described in case (1) is considered (Fig. 9a). Suppose that at wind speed V_w (12 m/s),   the   generator's   torque T_e and   turbine's torque T_m coincide at the optimum operating (point A) as seen from Fig. 15, when the wind speed changes to V_w (14 m/s) at $t = 2$ s, T_m changes abruptly and moves to point B, as shown in Fig. 16, where T_m is 194.8 N m. However, the change of the rotor speed is impeded by rotor inertia. Since the generator's torque T_e is governed by the rotor speed (as seen by the red curve in Fig. 15), the electromagnetic torque is slightly delayed. As the generator speed increases due torque difference $T_m - T_e$, the torque T_e increases, refer to the optimum torque curve. On the other hand, the turbine's torque drops with increasing the generator speed so that T_m and T_e eventually reach the same value at the point C. This is the maximum power point at wind speed V_w(14 m/s).

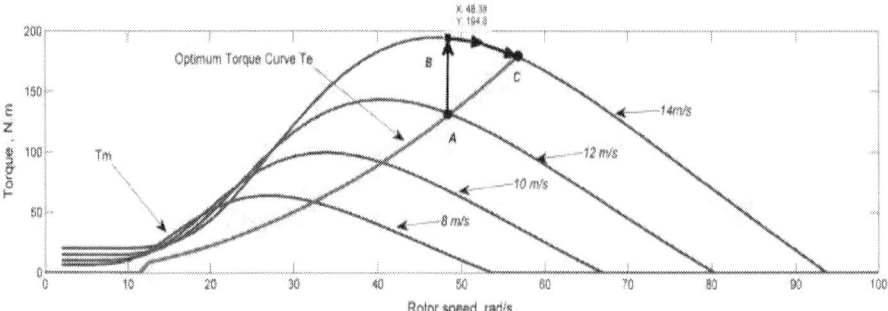

Figure 15. Torque versus generator speed characteristic.

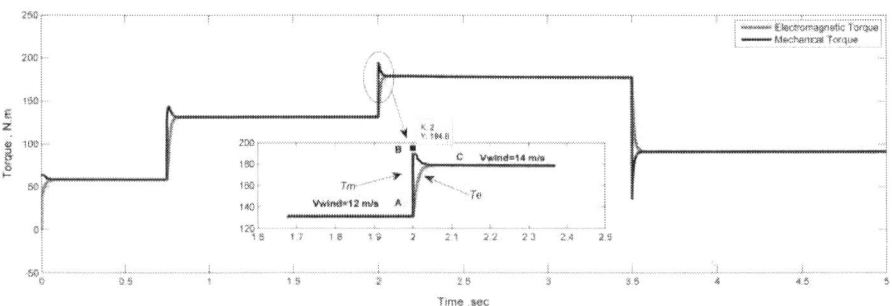

Figure 16. Response of wind turbine and generator torques.

FLICKER EMISSION OF PMSG-BASED GRID CONNECTED WIND TURBINES

Variable-speed wind turbines have shown better performance related to voltage fluctuation in comparison to fixed-speed wind turbines [10] and [13]. The reason is that variable-speed operation of the rotor has the advantage that the sharp power variations are not transmitted to the grid due to the presence of interfacing converters. In IEC 61400-21, the contribution of wind turbines to voltage fluctuations is divided into continuous operation and switching operation. This paper focuses only on flicker emission during continuous operation.

Flicker is defined as "an impression of unsteadiness of visual sensation induced by a light stimulus, whose luminance or spectral distribution fluctuates with time" [14]. Flicker causes consumer annoyance and complaint. Furthermore, it becomes a limiting factor for integrating wind turbines into weak grids, and even into relatively strong grids, where the wind power penetration levels are high.

A wind turbine with PMSG is integrated to external power system represented by constant voltage source connected in series with its Thevenin's equivalent impedance as in Fig. 17.

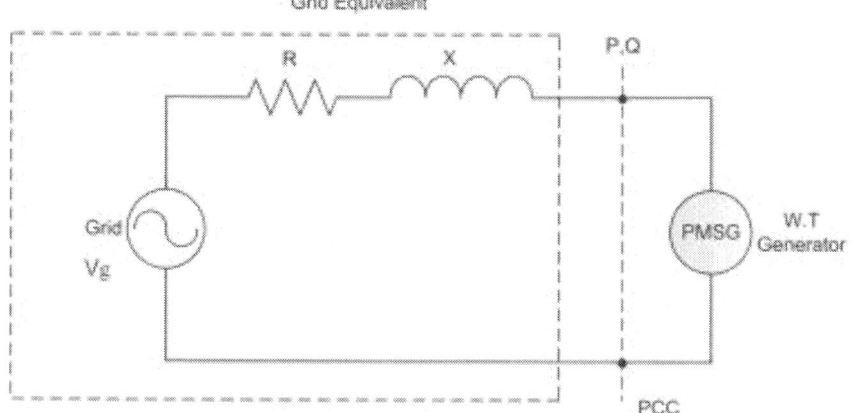

Figure 17. Simplified diagram of a grid-connected wind turbine.

In Fig. 17, the generator represents the wind turbine which is connected to the grid at the PCC. V_g is the voltage of the grid (V) and is assumed to 1 pu, R and X are the resistance and reactance of grid impedance (Ω), P_g and Q_g are the active (W) and reactive (Var) power flow produced by the wind turbine respectively. Since the grid voltage is constant, the voltage variation at the PCC is approximately calculated with the following formula[11], [12] and [14]:

$$\Delta V = \frac{P_g R + Q_g X}{V_g} \tag{18}$$

The fluctuations in the active or reactive power produced by the wind turbine result in voltage fluctuations and flicker at the PCC.

The level of flicker is quantified by the short-term flicker severity P_{st}, which is normally measured according to IEC standard IEC 61 000-4-15 [24], a flickermeter model is built to calculate the short-term flicker severity P_{st} which is calculated on the basis of voltage fluctuation of the PCC. More details about flickermeter description is presented in [8], [9],[14], [24] and [25].The flicker limits for low, medium, and extra high voltage networks are listed in Table 1 [24], [26] and [27].

Table 1. Flicker compatibility and planning levels.

Voltage level	Compatibility levels	Planning levels	
	LV and MV	MV	HV and EHV
P_{st}	1	0.9	0.8

FLICKER MINIMIZATION OF PMSG-BASED GRID-CONNECTED WIND TURBINES

Even though variable speed wind turbines have good performance with respect to flicker emission, flicker mitigation becomes necessary as the wind power penetration level increases. As mentioned before, the variable speed wind turbine with PMSG is capable of controlling the reactive power at the PCC. Normally, the output reactive power of the wind turbine is controlled at zero to maintain unity power factor. It is possible that the wind turbine output reactive power can be regulated by the grid-side converter control according to the output active power. In this approach, the grid-side converter behaves as a STATCOM at the PCC.

In this section, the voltage at the PCC is regulated by tuning the reactive power injection which can be realized by adding voltage regulator to the current loop of grid side converter control, as illustrated in Fig. 18. The reactive power adjustment can be used to minimize the voltage fluctuation caused by the active power flow.

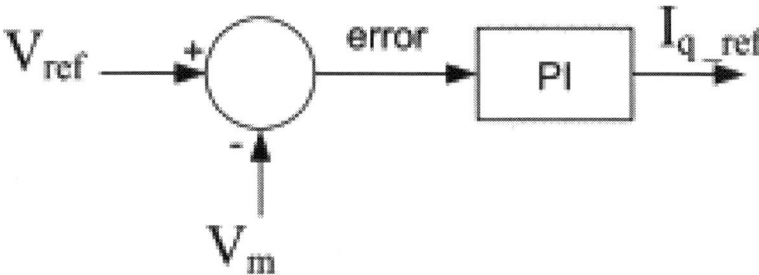

Figure 18. Voltage regulator (VR) control scheme.

The voltage regulator compares the reference voltage (which is 1 pu) with the measured voltage value and give the error signal to PI controller that is tuned to regulate the reactive power by I_{qref}. Thus keeps the voltage at a constant value which reduces voltage fluctuations and flicker emission as well.

SIMULATION RESULTS

Flicker emission of grid-connected wind turbines depends on several factors, such as grid parameters (e.g. fault level, grid impedance angle) site parameters (e.g. mean wind speed, turbulence intensity) and wind turbine configuration [8], [9], [10], [12],[13] and [15]. In the following cases, the concerned factors are to be changed while the other parameters are kept constant according to the base case of flicker simulation described in the appendix.

Fault Level
The impact of grid fault level (short circuit ratio) on the flicker emission is shown in Fig. 19. It can be seen that there is an inverse proportional relationship between P_{st} and fault level. In other words, the wind turbine would produce greater flicker in weak grids and vice versa. This simulation test is carried out with two cases of grid impedance angles, 45° and 75°. The mean wind speed at hub level equals12 m/s and site turbulence is 10%.

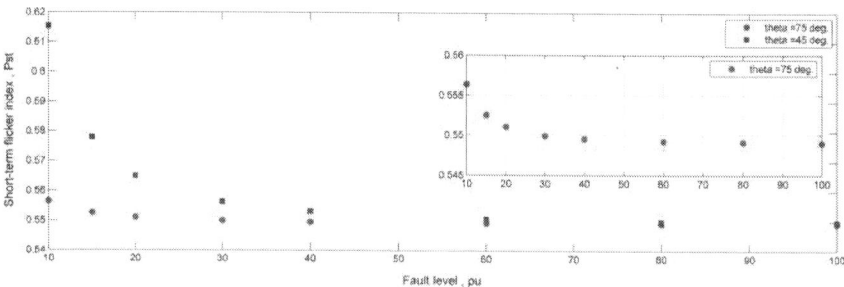

Figure 19. Variation of P_{st} with the grid fault level.

X/R Ratio of the Grid Impedance

The X/R ratio of the gird impedance is studied in terms of the impedance angle, $\Psi_K = \tan^{-1}(X/R)$. Fig. 20 depicts the effect of grid angle on flicker emission level. The mean wind speed at hub level is 12 m/s and site turbulence is 10%. Based on Fig. 20, at lower fault level (10 pu), greater flicker emission is induced compared to the values given with 100 pu fault level. At certain grid impedance angle, voltage variations caused by active power fluctuations could be cancelled by reactive power flow. The grid impedance angle Ψ_K can be defined in the following equation:

$$\tan \Psi_k = \frac{X}{R}$$

(19)

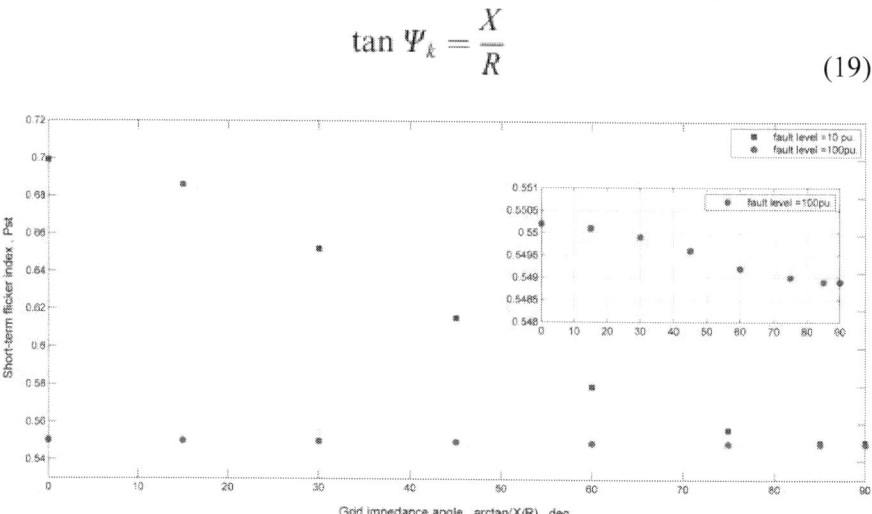

Figure 20. Variation of P_{st} with the grid impedance angle.

The wind turbine power factor angle Ψ is:

$$\tan \Psi = \frac{Q_g}{P_g}$$

(20)

Eq. (18) can be rewritten as follows

$$|\Delta V| = \frac{P_g R(1 + \tan \Psi_k \cdot \tan \Psi)}{V_g}$$

(21)

$$|\Delta V| = \frac{P_g R \cos(\Psi - \Psi_k)}{V_g \cos \Psi \cdot \cos \Psi_k}$$

(22)

From (21), when the difference between the grid impedance angle Ψ_K and the wind turbine power factor angle Ψ approaches 90°, $\Psi_K - \Psi = 90°$, the flicker emission is minimized. Normally the variable speed wind turbine with PMSG is controlled to operate at unity power factor [12], which means no reactive power is injected into or drawn from the grid, i.e., $\Psi = 0°$. To realize minimum voltage variation at the PCC, $\Psi_k = 90°$, i.e., $X \gg R$. Hence, for resistive grids (lower X/R ratios), the risk of flicker emission increases. As a result, the minimum value of flicker occurred at $\Psi_K = 90°$, which is clearly recognized in Fig. 20.

Mean Wind Speed

The variation of short-term flicker severity P_{st} with mean wind speed for the variable speed wind turbine with PMSG, is illustrated in Fig. 21. As shown, in low wind speeds (less than 10 m/s), the P_{st} value is very low due to a small output power so power fluctuations in the low wind region are low and therefore the resulting in small voltage fluctuation. Then the P_{st} value increases with an approximate linear relation with the mean wind speed until it reaches the rated wind speed. The voltage flicker decreases after the rated wind speed due to pitch control activation that can significantly limit the power output and smooth out the turbulence-induced fluctuations. This simulation is conducted with two cases of grid impedance angles; 45° and 75°. The mean wind speed at hub level is maintained at 12 m/s and site turbulence is 10%.

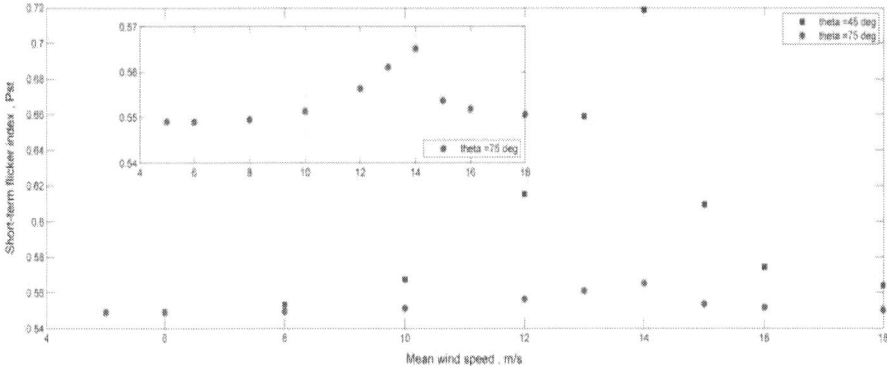

Figure 21. Variation of P_{st} with mean wind speed for $\Psi_K = 45°$ and $75°$.

Turbulence Intensity

In this case, the mean wind speed at hub level is maintained at 12 m/s, fault level is 10 pu, and the grid impedance angle is 45°. Fig. 22 shows the variation of P_{st} with the turbulence intensity. As the turbulence intensity increases, the wind profile changes significantly as in Fig. 23, which results in a large variation of output power. As consequence, the flicker emission increases with increase of the wind turbulence.

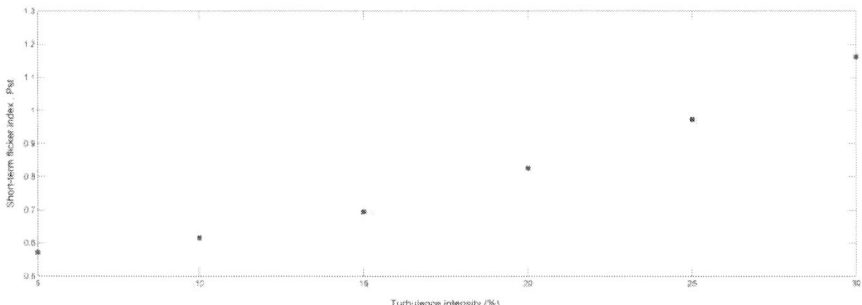

Figure 22. Variation of P_{st} with turbulence intensity for $\Psi_K = 45°$.

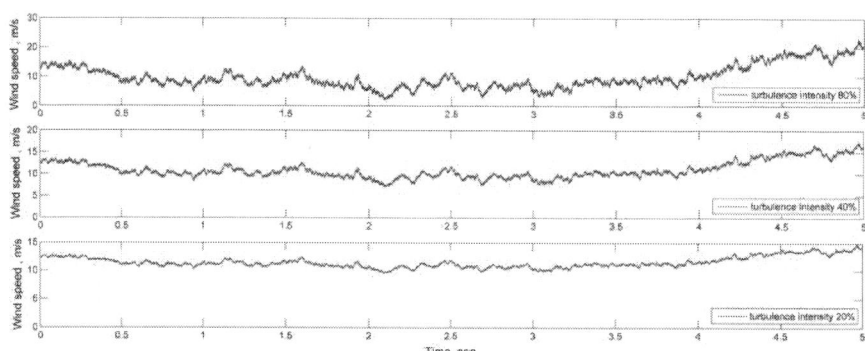

Figure 23. The wind speeds for different turbulence intensity values (v = 12 m/s).

The Impact of Voltage Regulator Loop

In this subsection, the impact of voltage regulator loop is highlighted by studying two cases depicting the system operation with and without loop activation. In the first case, described with different tracked signals in Fig. 24, the loop was not activated. Fig. 24a represents wind speed variation with 12 m/s and 10% mean value and turbulence intensity respectively. As seen in Fig. 24c, the output reactive power of the wind turbine is controlled to zero (unity power factor operation). From Fig. 24d, it can be seen that there are voltage fluctuations at the PCC, which is responsible of flicker emission phenomenon.

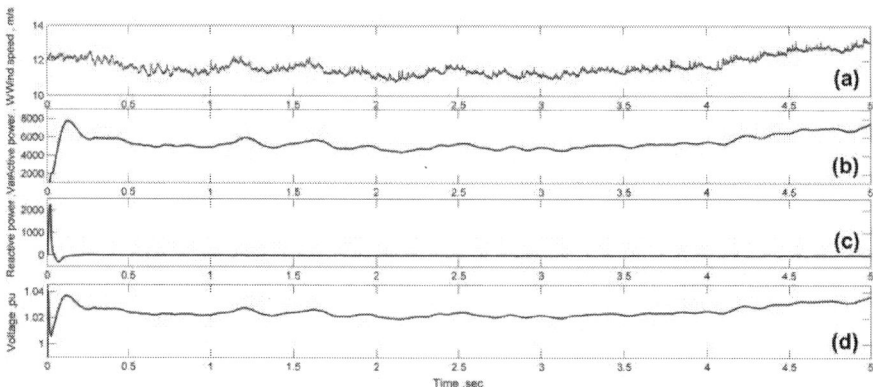

Figure 24. Performance of W.T based PMSG under normal operation without voltage regulation: (a) Wind speed; (b) Wind turbine output active power; (c) Wind turbine output reactive power; (d) Voltage at the PCC.

In the second case, described in Fig. 25, the voltage regulator loop is activated and the unity power operation is disabled. As seen in Fig. 25b, the voltage is highly smoothed and no voltage fluctuations can be detected. The reactive power (Fig. 25a) is varied and, of course, unity power operation is violated here. The flicker emissions for the two cases under study are depicted in Fig. 26 with different grid impedance angle. Further, Fig. 27presents flicker emissions for the two cases under study with different mean wind speeds.

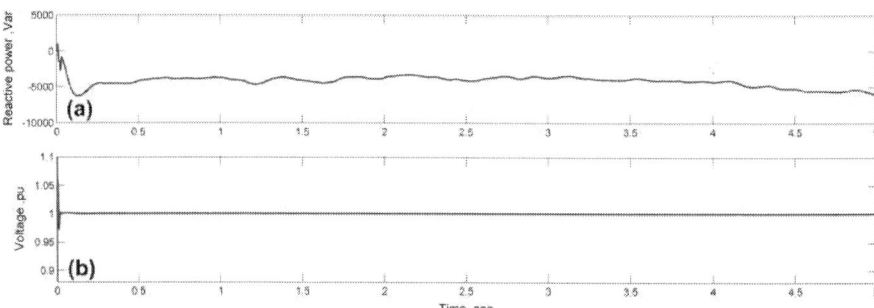

Figure 25. Performance of W.T based PMSG under voltage regulation: (a) Output reactive power; (b) output voltage at PCC.

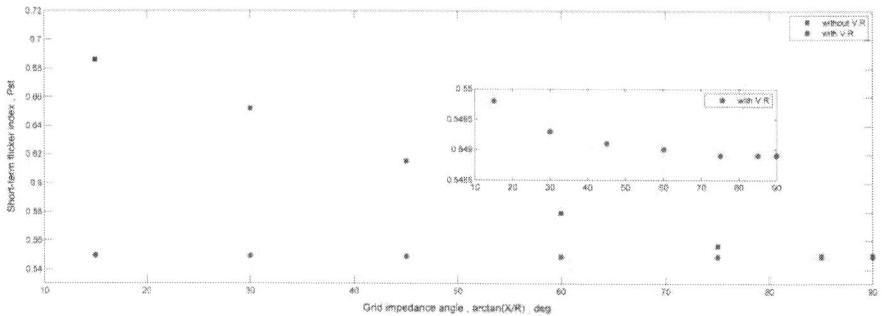

Figure 26. Variation of P_{st} with the grid impedance angle with and without VR.

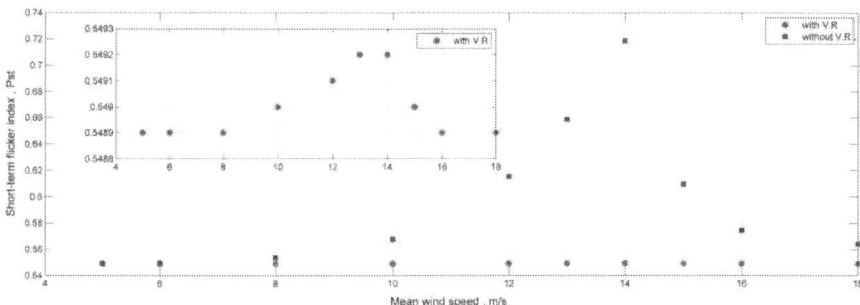

Figure 27. Variation of P_{st} with mean wind speed for $\Psi_K = 45°$ with and without regulation.

From Fig. 26, it can be concluded that the flicker level with VR loop is reduced and the effect of VR loop is more significant at low grid impedance angle.

As shown in Fig. 27, the flicker emission reduces when VR is activated. The effect of VR is significant in the linear region of wind turbine power curve (between the cut in and rated wind speeds). Beyond the rated wind speed, the pitch control is also affecting the flicker emission [15].

CONCLUSION

This paper presents comprehensive modeling of direct-driven PMSG-based grid-connected wind turbines along with the control schemes of the interfacing converters. In this configuration, two different control schemes are designed for the generator and grid side converters. Under different wind speed conditions, MPPT is provided via the developed control system of the generator side converter, while decoupled active and reactive power control is performed by the means of the grid side converter control. A voltage regulation loop is embedded in the control scheme of the grid-side converter to reduce flicker emission. Matlab/Simulink environment is conducted to examine the performance of the system.

Based on the developed wind turbine model, the flicker emission of variable speed wind turbines with PMSG during continuous operation is investigated. Simulation results show that the wind parameters and the grid parameters have significant effects on the flicker emission of the variable speed wind turbine with PMSG. The flicker level increases with increasing in turbulence intensity and mean wind speed however, it decreases with the increase of the grid impedance angle and fault level. The pitch control reduces the flicker level at higher wind speeds. Simulation results show the voltage regulator loop has a positive effect on flicker minimization during continuous operation of grid connected wind turbines. The paper discussion and results confirm the robustness and feasibility of .the developed control schemes of the generator and grid side converters.

Appendix A

Specification of wind turbine [17]

$$\frac{1}{\lambda_i} = \frac{1}{\lambda + 0.08\,\beta} - \frac{0.035}{\beta^3 + 1} \qquad \text{(A.1)}$$

The coefficients c_1 to c_6:

$c_1 = 0.5176$;	$c_2 = 116$;	$c_3 = 0.4$;
$c_4 = 5$;	$c_5 = 21$;	$c_6 = 0.0068$;
Wind turbine blade length:	$r = 2$ m;	
Air density:	$\rho = 1.225$ kg m^3;	
Optimal tip speed ratio:	$\lambda_{opt} = 8.1$;	
Maximum power coefficient:	$C_{p\_max} = 0.48$;	

Appendix B

Permanent magnet synchronous generator parameter

Stator resistance:	$R_s = 0.00829\ \Omega$;
Stator direct inductance:	$L_d = 0.174$ mH;
Stator quadrature inductance:	$L_q = 0.174$ mH;
Permanent magnet flux:	$\phi = 0.071$ wb;
No. of pole pairs:	$P = 6$ pair pole;
Inertia of the whole system:	$J = 0.089$ kg m²;
Friction factor:	$B = 0.005$ N m;

Appendix C

DC bus and grid parameters

dc-Link voltage:	$V_{DC} = 750$ V;
Capacitance of the dc link:	$C = 6000\ \mu F$;
Grid frequency:	$F = 60$ Hz;
Grid resistance:	$R_g = 0.02\ \Omega$;
Grid inductance:	$L_g = 0.05$ m;

Appendix D

Base case for flicker simulation

Fault level:	10 pu;
X/R of the grid impedance:	45 deg;
Mean wind speed V_{wind}:	12 m/s;
Turbulence intensity:	10%;

Appendix E

PI parameter

Generator side converter current loop:
$$K_p = 0.1;$$
$$K_i = 100;$$
Grid side converter current loop:
$$K_p = 2.5;$$
$$K_i = 500;$$
dc-link loop:-
$$K_p = 0.002;$$
$$K_i = 0.05;$$
VR loop:-
$$K_p = 0.055;$$
$$K_i = 1500.$$

REFERENCES

1. Ackermann T. Wind power in power systems. John Willey and Sons; 2005.
2. Tan K, Islam S. Optimum control strategies in energy conversion of PMSG wind turbine system without mechanical sensors. IEEE Trans Energy Convers 2004;19(2).
3. Baroudi JA, Dinavahi V, Knight AM. A review of power converter topologies for wind generators. Renew Energy 2007;32(January):2369–85.
4. Chinchilia M, Arnaltes S, Burgos J. Control of permanent-magnet generator applied to variable-speed wind-energy systems connected to the grid. IEEE Trans Energy Convers 2006;21(1).
5. Li S, Haskew TA, Xu L. Conventional and novel control designs for direct driven PMSG wind turbines. Electr Power Syst Res 2009;80(October):328–38.
6. Qiao W, Qu L, Harely RG. Control of IPM synchronous generator for maximum wind power generation considering magnetic saturation. IEEE Trans Ind Applic 2009;45(3).

7. Polinder H, der Pijl FFAV, Vilder Gert-Jan de, Tavner PJ. Comparison of direct-drive and geared generator concept for wind turbines. IEEE Trans Energy Convers 2006;21(3).

8. El-Tamaly HH, Wahab MAA, Kasem AH. Simulation of directly grid-connected wind turbines for voltage fluctuation evaluation. Int J Appl Eng Res 2007;2(1):15–30.

9. El-Tamaly HH, Wahab MAA, Kasem Ali H. Voltage fluctuation produced from wind turbines directly connected to grid. In: 10th International middle east power systems Conf., Mepcon'10th, Port-Said, Egypt, 13–15 December 2005.

10. Larsson A. Flicker emission of wind turbines during continuous operation. IEEE Trans Energy Convers 2002;17(1):114–8.

11. Hu W, Chen Z, Wang Y, Wang Z. Flicker mitigation by active power control of variable-speed wind turbines with full-scale back-to-back power converters. IEEE Trans Energy Convers 2009;24(3).

12. Sun T, Chen Z, Blaabjerg F. Flicker study on variable speed wind turbines with doubly fed induction generators. IEEE Trans Energy Convers 2005;20(4):896–905.

13. Papadopoulos MP, Papathanassiou SA, Tentzerakis ST, Boulaxis NG. Investigation of the flicker emission by grid connected wind turbines. In: Proc 8th Int Conf harmonics quality of power, vol. 2. Athens, Greece, October 14–16, 1998, p. 1152–7.

14. Kasem Ali H, El-Saadany Ehab F, El-Tamaly Hassan H, Wahab Mohamed AA. Power ramp rate control, voltage regulation and flicker mitigation of grid directly connected wind turbines. IET Renew Power Gener 2010;4(3):261–71.

15. Hu W, Chen Z, Wang Y, Wang Z. Flicker study on variable speed wind turbines with permanent magnet synchronous generator. In: 13th International power electronics and motion control conf., EPE-PEMC 13th, 2008, p. 2325–30.

16. Chen Z, Guerrero JM, Blaabjerg F. A review of the state of the art of power electronics for wind turbines. IEEE Trans Energy Convers 2009;24(8).

17. Heier S. Grid integration of wind energy conversion systems. John Wiley and Sons Ltd.; 1998, ISBN 0-471-97143-X.

18. Iov F, Hansen AD, Sorensen P, Blaabjerg F. Wind turbine blackest in Matlab/Simulink, Research Project, Institute of Energy Technology, Alborg University, March 2004.

19. Haque ME, Negnevistsky M, Muttaaqi KM. A naval control strategy for a variable speed wind turbine with a permanent magnet synchronous generator. IEEE Trans Ind Applic 2010;46(1).

20. Kasem AH, El-Saadany EF, El-Tamaly HH, Wahab MAA. An improved fault ride-through strategy for doubly fed induction generator-based wind turbines. IET Renew Power Gener 2008;2(4):201–14.
21. Muyeen SM, Takahashi R, Murata T, Tamura J. A variable speed wind turbine control strategy to meet wind farm grid code requirements. IEEE Trans Power Syst 2010;25(1).
22. Mehrzad D, Luque J, Quenca MC. Vector control of PMSG for grid connected wind turbine application, Project for Master Thesis, Institute of Energy Technology, Alborg University, Spring; 2009.
23. Morimoto S, Nakamura T, Takeda Y. Power maximization control of variable-speed wind generation system using permanent magnet synchronous generator. Electr Eng Jpn 2005;150(2), John Wiley & Sons.
24. IEC61000-4-15, Electromagnetic Compatibility (EMC) – Part 4: testing and measurements techniques – Section 15: Flickermeter, functional and design specifications, November 1997.
25. Ruiz J, Gutierrez JJ, Lazkano A, Ruiz de Gauna S. A review of flicker severity assessment by the IEC flickermeter. IEEE Trans Instrum Measur 2010;59(8).
26. IEC 61000-3-7 Technical Report Type 3: assessment of emission limits for fluctuation loads in MV and HV power systems Basic EMC Publication, 1996–10.
27. IEEE 1453-2004. Recommended practice for measurement and limits of voltage fluctuations and associated light flicker on AC power systems; 2004.

CITATION

Ali H. Kasem Alaboudy, Ahmed A. Daoud, Sobhy S. Desouky, Ahmed A. Salem, Converter controls and flicker study of PMSG-based grid connected wind turbines, Ain Shams Engineering Journal, Volume 4, Issue 1, March 2013, Pages 75-91, ISSN 2090-4479, http://dx.doi.org/10.1016/j.asej.2012.06.002.

CHAPTER 9

A Virtual Synchronous Machine Implementation for Distributed Control of Power Converters in Smartgrids

Salvatore D'Arco[1], Jon Are Suula[2], Olav B. Fosso[2]

[1]SINTEF Energy Research, 7465 Trondheim, Norway
[2]Department of Electric Power Engineering, Norwegian University of Science and Technology, 7495 Trondheim, Norway

ABSTRACT

The ongoing evolution of the power system towards a "SmartGrid" implies a dominant role of power electronic converters, but poses strict requirements on their control strategies to preserve stability and controllability. In this perspective, the definition of decentralized control schemes for power converters that can provide grid support and allow for seamless transition between grid-connected or islanded operation is critical. Since these features can already be provided by synchronous generators, the concept of Virtual Synchronous Machines (VSMs) can be a suitable approach for controlling power electronics converters. This paper starts with a discussion of the general features offered by the VSM concept in the context of SmartGrids. A specific VSM implementation is then presented in detail together with its mathematical model. The intended emulation of the synchronous machine characteristics is illustrated by numerical simulations. Finally, stability is assessed by analysing the eigenvalues of a small-signal model and their parametric sensitivities.

INTRODUCTION

The increasing penetration of power generation from renewable energy sources and the transition from a centralized power production model to distributed generation are expected to pose serious challenges to the development and operation of future power systems. This tendency is a strong motivation behind the paradigm shift from the traditional power system architecture towards an approach ensuring more flexibility and coordination between the generation units and loads that is promised by "SmartGrids"[1]. At the same time, the share of the electric power transferred through the power system which is processed by at least one power electronic conversion stage in the path from primary energy conversion to final consumption is continuously increasing. Already in 2007, it was estimated that this share would reach 80% around 2015 [2], and even if the development has been slightly slower, such a high share of power electronic conversion is expected to be exceeded during the coming years. Thus, power electronic control will have a crucial role in the emerging SmartGrid scenario, as the presence of power converters in the power system and their impact on global stability and controllability continues to increase.

Although the ongoing SmartGrid developments point towards an increasing level of communication and integration between various elements of the power system, distributed architectures with local primary control of converters combined with centralized secondary control seem to be an appropriate approach for optimizing steady-state operation while ensuring immediate response to transient events. Thus, converter units should be able to react autonomously to abrupt changes in the power system operating conditions, while complying on a longer time scale with the set-points and service requirements requested by the system operator through external communication.

In classical power systems, the Synchronous Machine (SM) with speed governor and excitation control offers favourable features to support the system operation within a distributed control scheme. Indeed, SMs contribute to the system damping through their inertia, participate in the primary frequency regulation through the droop response of the speed controller, and provide local control of voltage or reactive power flow. These capabilities, and especially the inertial and damping response common to all SMs, are not inherently offered by the power electronics interfaces commonly adopted for the integration of renewable energy sources. A distributed model for production and local control is also opening the possibility of islanded operation, which is inherently feasible with one or more controllable SMs in the islanded area. Such islanding operation is usually more complex to achieve with power converter interfaces designed for integration with a large-scale power system.

Power from many traditional large-scale generation facilities is currently being replaced by distributed generation capacity from wind power and photovoltaics. The traditional control structures implemented in the power converters for these applications rely on the synchronization to a stable grid frequency supported by large rotating inertias and are not inherently suitable in a SmartGrid context. Thus, from an implementation perspective, significant research efforts are still devoted towards development of control schemes for power electronic converters explicitly conceived to address the conditions emerging in future SmartGrids. Given the inherent benefits of the SMs outlined above, a captivating approach is the control of power electronic converters to replicate the most essential properties of the SM and by that gain equivalent features from a functional point of view. Thus, several alternatives for providing auxiliary services like reactive power control, damping of oscillations and emulation of rotating inertia with power electronic converters have been proposed [3], [4], [5], [6], [7] and [8]. Some of these control strategies are explicitly designed to mimic the dynamic response of the traditional SM, and can therefore be classified in broad terms as Virtual Synchronous Machines (VSM).

During the last decade, several concepts for VSMs have been presented with different names and different practical implementations [4], [8], [9], [10], [11] and [12]. The first review studies providing an overview of implementations have been recently published in[10] and [13], with an attempt to define a classification framework presented in [10]. The review in [10] also highlights how some implementations offer only partially the benefits of the SMs while only a few can ensure features as island operation of single or multiple units.
Most previous studies of VSM-based control strategies have presented particular implementation schemes which have been verified by time-domain simulations and/or laboratory experiments. A first study that included detailed modelling and small-signal stability of a particular VSM implementation was presented in [14]. However, this model was mainly developed for tuning of the converter control loops and did not consider the primary power-frequency control or the dynamics of the grid frequency detection needed to ensure an implementation of the VSM damping effect that adapts to variations in the grid frequency. A VSM system model addressing also these issues was recently presented in [15].

This paper includes a comprehensive treatment of a particular VSM implementation, starting from a discussion of the comparative advantages offered by the VSM concept in the context of SmartGrids in abstract terms. Then a brief overview of the VSM development status is offered, with the purpose of identifying general preferences for selecting specific implementations for future SmartGrid applications. The selected implementation is based on an internal representation of the SM inertia and damping behaviour through a reduced order swing equation, together with cascaded voltage and current controllers for operating a Voltage Source Converter (VSC), based on the general scheme from [15]. The paper derives step-by-step a detailed nonlinear mathematical model for this VSM implementation, and a corresponding small signal model in order to apply linear analysis techniques to the system in the perspective of stability assessment and controller tuning. The effect of system parameters on the poles of the linearized system model is also analyzed by calculating the parametric sensitivities of the system eigenvalues. The features and performance of the investigated VSM and its linearized small-signal model is verified with reference to a few selected cases by numerical simulations.

APPLICATION OF VIRTUAL SYNCHRONOUS MACHINES IN THE SMARTGRID CONTEXT

Power generation from distributed renewable energy sources like wind and photovoltaic power plants is usually connected to the power grid through actively controlled power electronic converters, and similar interfaces are applied for energy storage systems and an increasing share of controllable loads. The conventional scheme for such grid connected power converters is based on current controlled Voltage Source Converters (VSCs), which are synchronized to the measured grid voltage through a Phase Locked Loop (PLL) [16]. This approach usually requires a relatively strong grid with the presence of units that can maintain and stabilize the grid frequency and voltage. Even if auxiliary services like frequency and voltage support can be provided by current controlled VSCs, this functionality must be added through additional outer loop controllers which are not inherently applicable for operation in islanded mode [17]. Although this approach can be suitable for a relatively low penetration of grid connected converters, it does not seem sustainable for operation in a long term SmartGrid perspective with the expected dominant presence of power electronic conversion units and a high degree of flexibility in the network configurations.

Challenges for Power Converter Control in Future SmartGrids
In the last decade, several alternative concepts and approaches for control and operation of power converters distributed in the power system have

emerged. A noticeable example from the overall system operation point-of-view is the concept of Virtual Power Plants (VPPs) that aims to aggregate generation resources, energy storages and loads into clusters that can be controlled by the distribution system operator in a similar way as traditional power plants [18] and [19]. Such VPPs should coordinate the controllable units while ensuring supply to the uncontrolled loads in the system, but must also be able to supply auxiliary services like control of voltage or reactive power flow and support the frequency regulation of the system. In small isolated power systems, or in case parts of the distribution system should be able to operate in islanded mode, the generation units aggregated together in one VPP must also ensure a sufficient system inertia to keep the system stable while maintaining the power balance without large frequency deviations. The VPP concepts currently under development are capable of providing frequency-activated power support to the system within a few seconds, and can therefore ensure the steady-state power balance [20]. However, a faster response is required for ensuring an inertia-based power-frequency balance that will be able to keep the system stable in transient conditions. As mentioned, such a power-frequency response is a natural feature of traditional SMs which is not inherently present in the current controlled VSCs usually applied for integrating renewable sources to the grid. SMs offer also additional advantages as automatic synchronization and power sharing in response to changes in the operating conditions.

From these considerations it appears that a SmartGrid can represent a challenging environment for power converter control schemes, especially due to the possible large penetration of power electronics and the variability of the operating conditions. Indeed, a large penetration of converter-interfaced units will correspond to a lower level of physical inertia than in traditional power system dominated by large synchronous machines, and in certain conditions it can even be necessary to operate pure power electronic, inertia-less, systems. Considering that a flexible SmartGrid framework can result in more frequent reconfigurations of the power system, with corresponding variations in equivalent grid impedance and the possibility to operate parts of the system as groups of electrical islands, the variability of the conditions can introduce a further dimension of complexity. This can also lead to larger and more frequent transients and corresponding requirements for the control systems to maintain individual stable operation as well as contributing to the system stability in a wide range of operating conditions. The design of power electronic control schemes should cope with these challenges and preferably mitigate their effects.

One of the general characteristics of the emerging SmartGrid scenario is the presence of a communication infrastructure that can increase the volume of signal interaction between controllable units in the power system and facilitate their coordination. This can lead to a wide range of options for centralized control schemes, like the mentioned concept of VPPs, where references are determined by a centralized control unit and distributed to the individual converters. However, it should be noted that reducing the necessity of communication between the units, especially during transients, can increase the robustness of the system and reduce the risks in the event of temporary unavailability of the communication infrastructure. Thus, distributed control concepts where individual units can autonomously define their transient response based on local measurements are still relevant. Moreover, decentralized schemes where only steady-state references or set-points are distributed from a centralized system controller can lower the requirements in terms of bandwidth and latencies for the communication infrastructure, resulting in lower installation and operation costs. Multiple examples of possible solutions for decentralized control of power electronic converter can be found in the large literature on isolated MicroGrids [17], [21] and [22] although an exhaustive analysis is beyond the scope of this paper. However, not all of these schemes are suitable for SmartGrid applications where the converter are expected to operate most of the time in grid connected mode. It should also be mentioned that most control schemes that allow for both grid connected and stand-alone operation while also maintaining some decentralized control features, tend to be fairly complicated since transitions between these two modes usually require a reconfiguration of the control structure.

General Characteristics of VSMs

In the emerging SmartGrid context, the VSM concept can offer a basis for realizing flexible decentralized converter control schemes that can operate both in grid connected and islanded conditions, and that can almost seamlessly switch between the corresponding operating modes. Furthermore the inherent inertial characteristic of the VSM can provide services as frequency support and transient power sharing as primary control actions. These are indeed based only on local measurement and do not depend on external communications as in typical alternative schemes. Still, there is no conflict between this local controllability and the ability to operate in a hierarchical structure while following external references and set-points provided by a centralized controller for optimizing the system operation. Moreover, a further advantage of the VSM approach lies in its conceptual simplicity, due to the immediate and intuitive physical interpretation of its behaviour with analogy to the corresponding behaviour of a physical machine.

The dominant behaviour of SMs in terms of inertia response and damping can be modelled by the traditional swing equation [23]. Considering these general characteristics, several control strategies have been developed for allowing power electronic converters to provide synthetic or virtual inertia to the power system, and have been proposed for a variety of applications like for instance wind turbines, energy storage systems and HVDC transmission schemes [3], [4], [5], [6], [7], [8], [9], [10], [11], [12], [13], [14], [24], [25] and [26]. Some of these control methods provide a synthetic inertial response to variations in the grid frequency and only a few aims to explicitly replicate the features of the traditional SMs. However, emulation of the inertia and damping effects requires an energy buffer with sufficient capacity to represent the energy storage effect of the emulated rotating inertia available. Thus, the amount of virtual inertia that can be added to the system by a single VSM unit will be limited by the DC-side configuration and by the current rating of the converter.

The current state of the art on the VSM concept has been presented in [10] and [13]. These reviews identify that some of the implementations proposed as VSMs do not exploit the full potential of the concept because they still rely on a PLL for detecting the grid voltage phase angle, the grid frequency and its derivative, thus requiring the presence of rotating inertia in the grid. Other proposed implementations of the VSM concept are based on the simulation of an internal mathematical SM model inside the control system providing a voltage reference output for the PWM [9]. However, direct open loop PWM signal generation from these voltage references prevents the possibility to explicitly embed the limitations and controlled saturations of voltages and currents that are normally required as protective functions for safe operation of power electronic converters. These protective functions can be easily included in a cascaded control scheme [10], [17], [27] and [28] where the output from the VSM inertia emulation is used as reference for a voltage control loop cascaded with an internal current control loop. Numerically, this approach is sufficiently robust for practical implementations and will in the following be assumed as the reference VSM scheme, elaborated from previous studies in [14] and [15]. It can also be noted that the implementation of a VSM based on the swing equation providing references for operation of the converter, under certain conditions has been shown to be equivalent to the frequency-droop-based control strategies first developed for Uninterruptable Power Supply (UPS) systems and MicroGrids, as demonstrated in [10] and [29]. However, the interpretation of the parameters in a VSM approach seems to be simpler and more intuitive than the equivalent parameters in the commonly applied MicroGrid schemes and, thus, preferable.

MATHEMATICAL MODEL OF THE VSM REFERENCE IMPLEMENTATION

This section describes the control scheme for the selected VSM reference implementation, considering each functional block, and derives the corresponding mathematical model.

Control System Overview

An overview of the studied VSM configuration is shown in Fig. 1, where a VSC is connected to a grid through an LC filter. In the following, the switching effects of the VSC are neglected and an ideal average model is assumed for modelling the converter. Furthermore, no application-specific constraints of the DC-side of the VSC are considered and, thus, modelling and control of the energy source or storage on the DC side of the converter is not further discussed.

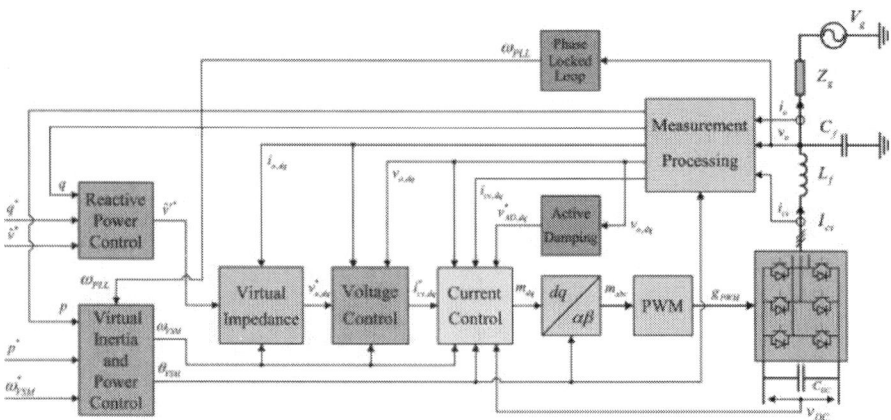

FIGURE 1. Overview of investigated system configuration and control structure for the Virtual Synchronous Machine.

The VSM-based power control with virtual inertia provides frequency and phase angle references ω_{VSM} and θ_{VSM} to the internal control loops for operating the VSC, while a reactive power controller provides the voltage amplitude reference $\hat{v}^{r*}$. Thus, the VSM inertia emulation and the reactive power controller appear as outer loops providing the references for the cascaded voltage and current controllers. A PLL detects the actual grid frequency, but this frequency is only used for implementing the damping term in the swing equation. Thus, the operation of the inner loop controllers does not rely on the PLL as in conventional VSC control systems, but only on the power-balance-based synchronization mechanism of the VSM inertia.

Modelling Conventions
In Fig. 1, upper case symbols represent physical values of the electrical circuit. The control system implementation and the modelling of the system are based on per unit quantities, denoted by lower case letters where the base values are defined from the apparent power rating and the rated peak value of the phase voltage [30].

The modelling, analysis and control of the electrical system is implemented in Synchronous Reference Frames (SRFs). The transformation from the stationary reference frame into the SRFs are based on the amplitude-invariant Park transformation, with the d-axis aligned with a voltage vector and the q-axis leading the d-axis by 90° [30]. Thus, the magnitude of current and voltage vectors at rated conditions is 1.0 pu. Whenever possible, SRF equations are presented in complex space vector notation as:

$$\mathbf{x} = x_d + j \cdot x_q \tag{1}$$

Thus, active and reactive powers can be expressed on complex or scalar form as:

$$p = Re(\mathbf{v} \cdot \mathbf{i}) = v_d \cdot i_d + v_q \cdot i_q$$

$$q = Im(\mathbf{v} \cdot \breve{\mathbf{i}}) = -v_d \cdot i_q + v_q \cdot i_d \tag{2}$$

The current directions indicated in Fig. 1 result in positive values for active and reactive powers flowing from the converter into the grid.

System Modelling

In the following sub-sections, the implementation of each functional block of the VSM-based control and the mathematical models of all system elements from Fig. 1 are presented as a basis for developing a non-linear model of the system. This system model will also be used to establish a linearized small-signal state-space representation.

VSM Inertia Emulation and Active Power Droop Control

The emulation of a rotating inertia and the power-balance based synchronization mechanisms of this virtual inertia is the main difference between the investigated VSM control structure and conventional control systems for VSCs. The VSM implementation investigated in this case is based on a conventional swing equation representing the inertia and damping of a traditional SM [10] and [14]. The swing equation used for the implementation is linearized with respect to the speed so that the acceleration of the inertia is determined by the power balance according to:

$$\frac{d\omega_{VSM}}{dt} = \frac{p^{r*}}{T_a} - \frac{p}{T_a} - \frac{p_d}{T_a} \tag{3}$$

In this equation, $pr*$ is the virtual mechanical input power, p is the measured electrical power flowing from the VSM into the grid, and pd is the damping power, while the mechanical time constant is defined as Ta (corresponding to $2H$ in a traditional SM). The per unit mechanical speed ω_{VSM} of the virtual inertia is then given by the integral of the power balance while the corresponding phase angle θ_{VSM} is given by the integral of the speed. A block diagram showing the implementation of the VSM swing equation is shown on the right in Fig. 2.

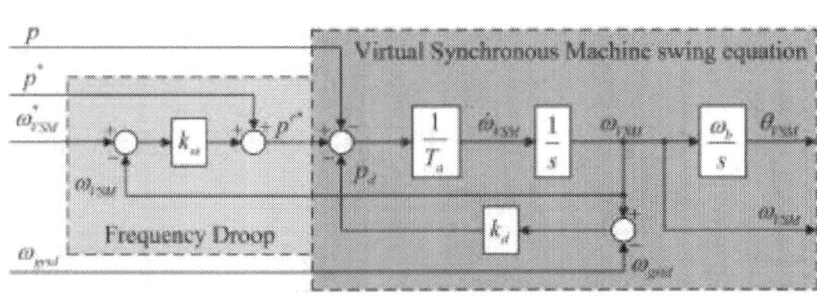

FIGURE 2. Virtual Synchronous Machine inertia emulation with power-frequency droop.

The VSM damping power pd, representing the damping effect of a traditional SM, is defined by the damping constant kd and the difference between the VSM speed and the actual grid frequency. Thus, an estimate of the actual grid frequency is needed for the VSM implementation. As indicated in the figure, the frequency estimate is in this case labelled as ωPLL and is provided by a PLL.

An external frequency droop, equivalent to the steady-state characteristics of the speed governor for a traditional synchronous machine, is included in the power control of the VSM as shown in the left part of Fig. 2. This power-frequency droop is characterized by the droop constant $k\omega$ acting on the difference between a frequency reference ω^*_{VSM} and the actual VSM speed ω_{VSM}. Thus, the virtual mechanical input power pr^* to the VSM swing equation is given by the sum of the external power reference set-point, p^*, and the frequency droop effect, as shown on the left of Fig. 2.

For modelling the VSM in a SRF, the phase angle of the VSM in grid connected mode should be constant under steady-state conditions and should correspond to the phase displacement between the virtual position of the VSM internal voltage and the position of the grid voltage vector. Since only the deviation of the VSM speed from the actual grid frequency should be modelled to achieve this, a new set of variables representing the speed deviation $\delta\omega VSM$ and the corresponding phase angle difference $\delta\theta VSM$ is introduced. Thus, the power balance of the VSM inertia can be expressed by (4), while the VSM phase displacement is defined by (5):

$$\frac{d\delta\omega_{VSM}}{dt} = \frac{p^*}{T_a} - \frac{p}{T_a} - \frac{k_d(\omega_{VSM} - \omega_{PLL})}{T_a} - \frac{k_\omega(\omega_{VSM} - \omega^*)}{T_a} \tag{4}$$

$$\frac{d\delta\theta_{VSM}}{dt} = \delta\omega_{VSM} \cdot \omega_b \tag{5}$$

Since the VSM speed in steady state will become equal to the grid frequency ωg, the frequency deviation $\delta\omega_{VSM}$ will return to zero under stable grid connected operation.

The actual per unit speed of the VSM shown in the block diagram of Fig. 2 can be expressed from the speed deviation $\delta\omega_{VSM}$ resulting from (4) and the grid frequency ωg as given by (6). The corresponding VSM phase angle θ_{VSM} is then defined by (7)

$$\omega_{VSM} = \delta\omega_{VSM} + \omega g \tag{6}$$

$$\frac{d\theta_{VSM}}{dt} = \omega_{VSM} \cdot \omega_b$$

(7)

The phase angle θVSM will then become a saw-tooth signal between 0 and 2π, which is the phase angle that will be used for the transformation between the rotating reference frame defined by the VSM inertia and the three-phase signals, as indicated in Fig. 1.

Reactive Power Droop Controller
The droop-based reactive power controller applied in this case is similar to the controllers commonly applied in microgrid systems [17] and [27]. The voltage amplitude reference $\hat{v}^{r*}$ used for the inner loop voltage and current control is then calculated by (8) where $\hat{v}^*$ is the external voltage amplitude reference and $q*$ is the reactive power reference. The gainkq is the reactive power droop gain acting on the difference between the reactive power reference and the filtered reactive power measurement qm. The state of the corresponding first order low pass filter applied in this case is defined by (9), where ωf is the cut-off frequency. A block diagram of the resulting control structure is shown in Fig. 3:

$$\hat{v}^{r*} = \hat{v}^* + k_q(q^* - q_m)$$

(8)

$$\frac{dq_m}{dt} = -\omega_f \cdot q_m + \omega_f \cdot q$$

(9)

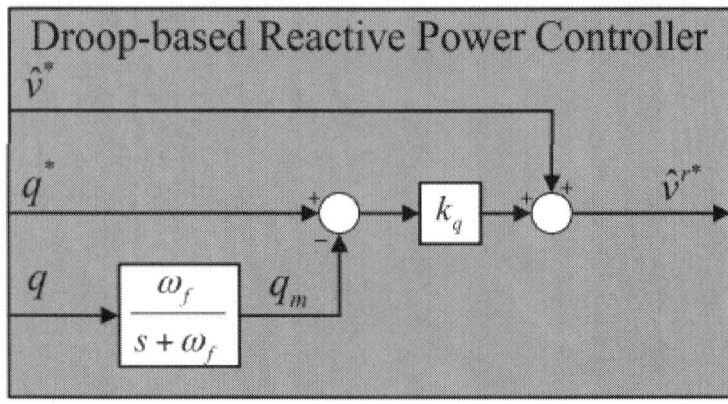

FIGURE 3. Reactive power droop controller.

Reference Frame Orientations

The synchronization of the VSM control system to the grid is based on the phase angle orientation of the virtual rotor of the VSM, and the phase angle θ_{VSM} is used in the transformations between the stationary reference frame and the VSM-oriented SRF. Thus, the power balance of the VSM swing equation will ensure the synchronization to the grid voltage without the need for a traditional PLL. Since the VSM-oriented SRF in steady state rotates with the same frequency as the grid voltage, this phase angle will be continuously increasing between 0 and 2π, as indicated in the vector diagram shown in Fig. 4. According to its definition, the phase angle $\delta\theta_{VSM}$ is instead representing the phase difference between the VSM SRF orientation and the rotating grid voltage vector, as also indicated in Fig. 4.

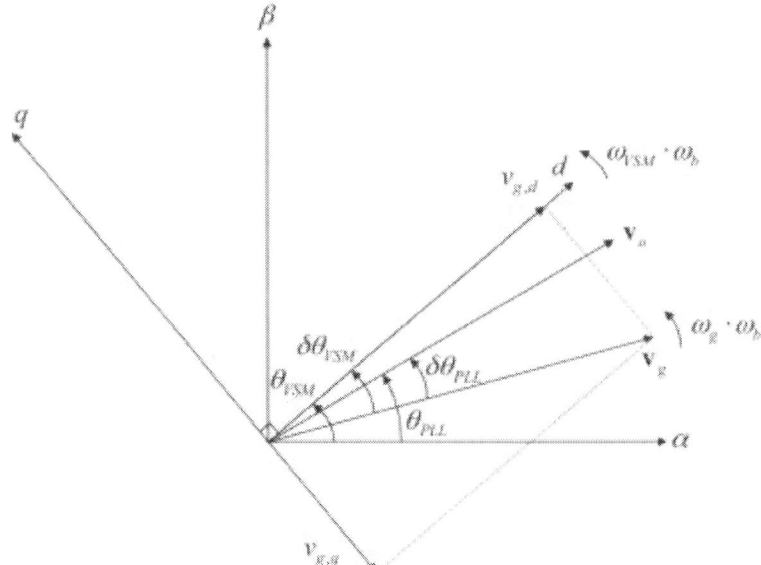

Figure 4. Vector diagram defining the SRF and voltage vector orientations.

The VSM-oriented SRF is used for both control and modelling of the system, and therefore, also the model of the electrical system will be represented in this reference frame. This has significant advantages for the modelling of the system, since multiple reference frame transformations between a local SRF for controller implementation and a global SRF for electrical system modelling can be avoided. Considering the amplitude of the equivalent grid voltage $\hat{V}_g$ to be known, the voltage vector **vg** in the VSM-oriented SRF can then be expressed by (10):

$$\mathbf{V_g} = \hat{v}_g e^{-j\delta\theta_{VSM}}$$

(10)

By the power-balance-based synchronization effect of the VSM swing equation, the control system defines its own reference frame orientation with respect to the grid voltage. In principle, no additional reference frames are needed to model the system from Fig. 1. However, since an estimate for the grid frequency is used to implement the VSM damping effect, a PLL operating on the measured voltage $\mathbf{v}o$ at the filter capacitors is implemented as part of the control system. Thus, this PLL will establish its own SRF aligned with the voltage vector $\mathbf{v}o$. The phase angle displacement of this PLL with respect to the grid voltage can then be defined as $\delta\theta_{PLL}$ in a similar way as for the phase angle displacement of the VSM. The detailed implementation of the PLL will be presented in the following sub-section, but the definition of its steady state phase displacement $\delta\theta_{PLL}$ with respect to the grid voltage, and the corresponding phase angle θ_{PLL} between the rotating PLL-oriented SRF and the stationary reference frame, is shown in Fig. 4.

According to the definitions indicated in Fig. 4, the phase angle between the VSM and PLL oriented SRFs will be defined by the difference between the VSM and PLL angles. For modelling of the PLL in its own reference frame, the voltage $\mathbf{v}o$ at the filter capacitors can be transformed from the VSM-oriented reference frame to the PLL-oriented reference frame by:

$$\mathbf{v}_0^{PLL} = \mathbf{v}_0^{VSM} e^{-j(\delta\theta_{PLL} - \delta\theta_{VSM})}$$

(11)

Phase Locked Loop

The Phase Locked Loop (PLL) applied in this case for tracking of the actual grid frequency is based on [31] and [32] and its structure is shown in Fig. 5. This PLL is using first order low-pass filters on the estimated d- and q-axis voltage components and an inverse tangent function to calculate the phase angle error of the PLL. This phase angle error e_{PLL} is the input to a PI controller tracking the frequency of the measured voltage. For the practical implementation, the estimated frequency ω_{PLL} is then integrated to obtain the estimate of the actual instantaneous phase angle θ_{PLL} used for transformation of the voltage measurements into the PLL-oriented SRF. For modelling of the PLL, the voltage vector $\mathbf{v}o$ in the VSM-oriented SRF must be transformed into the PLL-oriented SRF according to (11).

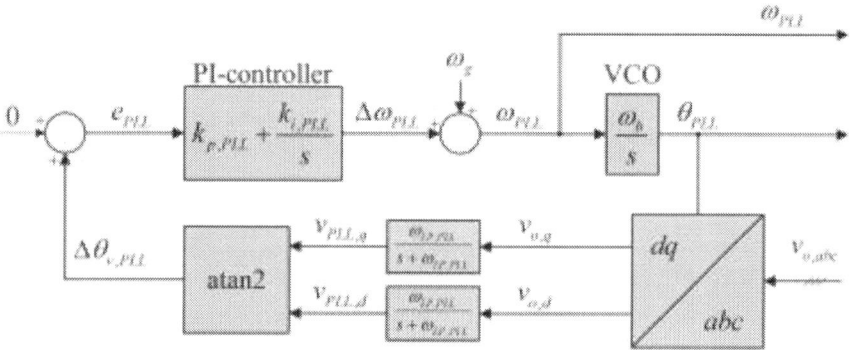

Figure 5. Phase locked loop.

The states of the applied first order low-pass filters in the PLL, defining the filtered voltage v_{PLL}, can be expressed by (12), where the cut-off frequency of the applied low pass filters is given by $\omega_{LP,\,PLL}$:

$$\frac{d\mathbf{v}_{PLL}}{dt} = -\omega_{LP.PLL} \cdot \mathbf{V}_{PLL} + \omega_{LP.PLL} \cdot \mathbf{V}_o e^{-j(\delta\theta_{PLL} - \delta\theta_{VSM})} \tag{12}$$

The integrator state ε_{PLL} of the PI controller can then be defined by:

$$\frac{d\varepsilon_{PLL}}{dt} = \tan^{-1}\left(\frac{V_{PLL,q}}{V_{PLL,d}}\right) \tag{13}$$

In the same way as explained for the SRF modelling of the VSM swing equation, a speed deviation $\delta\omega_{PLL}$ with respect to the grid frequency is defined for the PLL according to (14). The corresponding phase angle displacement, $\delta\theta_{PLL}$, of the PLL is then defined by (15):

$$\delta\omega_{PLL} = k_{p,PLL} \cdot \tan^{-1}\left(\frac{V_{PLL,q}}{V_{PLL,d}}\right) + k_{i,PLL} \cdot \varepsilon_{PLL} \tag{14}$$

$$\frac{d\delta\theta_{PLL}}{dt} = \delta\omega_{PLL} \cdot \omega_b \tag{15}$$

In accordance with the definitions introduced for the VSM swing equation, the actual per unit frequency ωPLL detected by the PLL is given by (16). The phase angle used in the implementation of the PLL, for transformation of the measured three-phase voltage measurements into the PLL-oriented SRF, is then defined by θPLL according to (17):

$$\omega_{PLL} = \delta\omega_{PLL} + \omega g \tag{16}$$

$$\frac{d\theta_{PLL}}{dt} = \omega_{PLL} \cdot \omega_b \tag{17}$$

Virtual Impedance and Voltage Controllers

As indicated in Fig. 1, the voltage amplitude reference $\hat{v}^{r*}$ resulting from the reactive power droop controller in Fig. 3 is passed through a virtual impedance before it is used as a reference for controlling the voltage vo at the filter capacitors. This virtual impedance can be considered as an emulation of the quasi-stationary characteristics of the synchronous impedance in a traditional SM. The virtual impedance will influence the steady-state and dynamic operation of the VSM, and it will be shown how it can be used to shape the dynamic characteristics of the system. Since power flowing through the virtual inductance will cause a phase angle displacement between the grid voltage and the virtual inertia position of the VSM, it will also reduce the sensitivity of the VSM to small disturbances in the grid. The influence from the virtual resistance rv and inductance lv on the capacitor voltage reference vector $\mathbf{V_o^*}$ is defined on basis of the current io according to[33] and [34]:

$$\mathbf{v}_o^* = \hat{v}^{r*} - (r_v + j \cdot \omega_{VSM} \cdot l_v) \cdot \mathbf{i_o} \tag{18}$$

The resulting d- and q -axis voltage components $v_{o,d}^*$ and $v_{o,q}^*$ are used directly as references for the decoupled SRF PI voltage controllers as shown in the left part of Fig. 6.

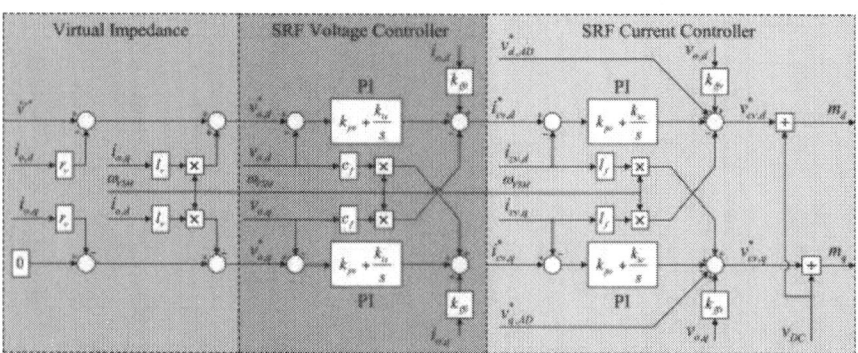

FIGURE 6. Virtual inductance, voltage control and current control.

The detailed structure of the SRF PI controllers for the filter capacitor voltage is shown in the middle of Fig. 6, and is producing the reference values $\mathbf{i}_{cv}^*$ for the converter currents[27]. These current references can be expressed by (19), where the PI controller gains are defined by kpv and kiv. A gain factor $kffi$ that can be set to 1 or 0 is used to enable or disable the feed-forward of measured currents flowing into the grid. It should also be noted that the decoupling terms of the voltage controller are based on the per unit speed of the VSM inertia as defined in (6). The states ξ are defined to represent the integrators of the PI voltage controllers as given by (20):

$$\mathbf{i}_{cv}^* = k_{pv}(\mathbf{v}_o^* - \mathbf{v}_o) + k_{iv}\xi + j \cdot c_1 \cdot \omega_{VSM} \cdot \mathbf{v}_o + k_{ffi} \cdot i_o \tag{19}$$

$$\frac{d\xi}{d} = \mathbf{v}_o^* - \mathbf{v}_o \tag{20}$$

The current references from the voltage controllers should be limited to avoid over-currents in case of voltage drops, fault conditions or other severe transients. This also implies that the voltage controllers must be protected from windup conditions in case the current references are saturated. However, the required limitations and anti-windup techniques for the investigated VSM scheme are similar to what is needed in conventional droop-based control schemes with cascaded SRF voltage and current controllers, as for instance discussed in [28]. Since these limitations are not influencing the dynamics of the control scheme within the normal operating range, further details will not be discussed here.

Current Controllers and Active Damping

The applied inner loop current controllers are conventional SRF PI controllers with decoupling terms [27] and [35], as shown in the right side part of Fig. 6. The output voltage reference from the controller is defined by (21), where the resulting voltage reference for the converter is denoted by $\mathbf{v}_{cv}^*$. The proportional and integral gains of the PI controller are defined by kpc and kic, and a gain factor $kffv$ is used to disable or enable the voltage feed-forward in the output of the current controllers. The states γ are defined to represent the integrators of the PI controllers according to (22):

$$\mathbf{v}_{cv}^* = k_{pc}(\mathbf{i}_{cv}^* - \mathbf{i}_{cv}) + k_{ic} \cdot \gamma + j \cdot l_1 \cdot \omega_{VSM} \cdot \mathbf{i}_{cv} + k_{ffv} \cdot \mathbf{v}_o - \mathbf{v}_{AD}^* \tag{21}$$

$$\frac{d\gamma}{dt} = \mathbf{i}^*_{cv} - \mathbf{i}_{cv}$$

(22)

In (21), the voltage reference for the converter also includes an active damping term $\mathbf{V}^*_{AD}$ designed for suppressing LC oscillations in the filter [36]. The implementation of the active damping algorithm is shown in Fig. 7, and is based on high pass filtering of the measured voltage vo, obtained from the difference between vo and the low pass filtered value of the same voltage. The resulting high pass filtered signal is then scaled by the gain kAD according to (23) and subtracted from the output of the current controllers to cancel detected oscillations in the capacitor voltages:

$$\mathbf{V}^*_{AD} = k_{AD}(\mathbf{V}_0 - \varphi)$$

(23)

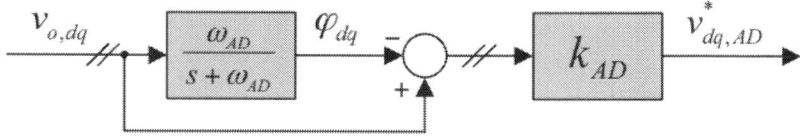

FIGURE 7. Implementation of active damping.

The corresponding internal states φ of the low pass filters used for the active damping are defined by (24), where ωAD is the cut-off frequency:

$$\frac{d\varphi}{dt} = \omega_{AD} \cdot \mathbf{V}_0 - \omega_{AD} \cdot \varphi$$

(24)

For the practical implementation of the VSC control system, the voltage reference $\mathbf{V}^*_{cv}$ resulting from the current controller and the active damping is divided by the measured DC-link voltage to result in the modulation index $\mathbf{m}$ as shown to the right of Fig. 6. Neglecting the switching operation of the converter and any delay due to the PWM implementation, the instantaneous average value of the per unit converter output voltage is given by the product of the modulation index and the actual DC-voltage. Under this assumption, the output converter voltage will be approximately equal to the voltage reference as summarized by (25)[37]:

$$\mathbf{m} = \frac{\mathbf{V}^*_{cv}}{v_{DC}}, \quad \mathbf{V}_{cv} = \mathbf{m} \cdot v_{DC} \rightarrow \mathbf{V}_{cv} \approx \mathbf{V}^*_{cv}$$

(25)

Thus, the AC side operation of the converter will be effectively decoupled from any dynamics in the DC voltage, and it is not necessary to further discuss or model the DC side of the converter for achieving an accurate representation of the dynamics on the AC side. It should be noted that the actual source or storage unit connected to the DC link of the converter might still impose restrictions on the allowable power exchange during various operating conditions. However, to maintain generality and avoid detailed discussion of particular application-specific limitations it will be assumed that the power requested from the AC side is always available at the DC link of the converter.

Electrical System Equations

The electrical system included in the model according to Fig. 1 consists of a set of filter inductors connected to the converter, a shunt capacitor bank representing the capacitance of the LC filter, and a Thévenin equivalent of the grid. This simple structure is assumed to achieve a simple model that mainly includes the dynamics of the converter control system and its interaction with the equivalent grid voltage. However, a more complex AC grid topology can be easily included in the model for both simulations and analysis. Considering an instantaneous average model of the converter, the SRF state space equations of the electrical system can be established as given by (26)[27] and [35]:

$$\frac{d\mathbf{i}_{cv}}{dt} = \frac{\omega_b}{l_f}\mathbf{v}_{cv} - \frac{\omega_b}{l_f}\mathbf{v}_o - \left(\frac{r_{l_f}\omega_b}{l_f} + j \cdot \omega_g \omega_b\right)\mathbf{i}_{cv}$$

$$\frac{d\mathbf{v}_o}{dt} = \frac{\omega_b}{c_f}\mathbf{i}_{cv} - \frac{\omega_b}{c_f}\mathbf{i}_g - j \cdot \omega_g \omega_b \cdot \mathbf{v}_o$$

$$\frac{d\mathbf{i}_o}{dt} = \frac{\omega_b}{l_g}\mathbf{v}_o - \frac{\omega_b}{l_g}\mathbf{v}_g - \left(\frac{r_g\omega_b}{l_g} + j \cdot \omega_g \omega_b\right)\mathbf{i}_o \tag{26}$$

In these equations $\mathbf{i}_{cv}$ is the filter inductor current, $\mathbf{v}_{cv}$ is the converter output voltage, $\mathbf{v}_o$ is the voltage at the filter capacitors, $\mathbf{i}_g$ is the current flowing into the grid equivalent and $\mathbf{v}_g$ is the grid equivalent voltage. The inductance and equivalent resistance of the filter inductor is given by l_f and r_{lf}, the filter capacitor is c_f, while the grid inductance and resistance are given by l_g and r_g. The per unit grid frequency is given by ω_g, while the base angular grid frequency is defined by ω_b. It should be noted that the state space model from (26) can represent the electrical system in any SRF, but in this case the system will always be modelled in the SRF defined by the VSM swing equation.

It should also be noted that the presented model only represents the case of grid-connected operation, while the investigated VSM scheme is inherently suitable for stand-alone operation. In this case the operational frequency will only be determined by the actual load in the system, the power-frequency droop gain and the power and frequency references for the VSM. Further details on modelling of the investigated VSM implementation in islanded operation, and corresponding analysis of the dynamic characteristics in stand-alone mode can be found in [38].

Non-linear System Model in Grid-connected Operation

All equations needed for detailed modelling of the VSM configuration in grid-connected operation have been presented in the previous sub-sections, and can be reduced to a model on state-space form with 19 distinct state variables and 6 input signals, with the state vector $\mathbf{x}$ and the input vector $\mathbf{u}$ defined by (27). The resulting non-linear state-space model of the overall system is given by (28):

$$\mathbf{x} = \begin{bmatrix} v_{o,d} & v_{o,q} & i_{cv,d} & i_{cv,q} & \gamma_d & \gamma_q & i_{o,d} & i_{o,q} & \varphi_d & \varphi_q \cdots \\ \cdots v_{PLL,d} & v_{PLL,q} & \varepsilon_{PLL} & \delta\theta_{VSM} & \xi_d & \xi_q & q_m & \delta\omega_{VSM} & \delta\theta_{PLL} \end{bmatrix}^T$$

$$\mathbf{u} = \begin{bmatrix} p^* & q^* & \hat{v}_g & \hat{v}^* & \omega^* & \omega_g \end{bmatrix}^T$$

$$(27)$$

$$\frac{dv_{o,d}}{dt} = \omega_b \omega_g v_{o,q} + \frac{\omega_b}{c_f} i_{cv,d} - \frac{\omega_b}{c_f} i_{o,d}$$

$$\frac{dv_{o,q}}{dt} = -\omega_b \omega_g v_{o,d} + \frac{\omega_b}{c_f} i_{cv,q} - \frac{\omega_b}{c_f} i_{o,q}$$

$$\frac{di_{cv,d}}{dt} = \frac{\omega_b(k_{ffv} - 1 - k_{AD} - k_{pc}k_{pv})}{l_f} v_{o,d} - \frac{\omega_b c_f k_{pc}}{l_f} \omega_g v_{o,q} - \frac{\omega_b(k_{pc} + r_f)}{l_f} i_{cv,d} + \frac{\omega_b k_{ic}}{l_f} \gamma_d + \frac{\omega_b k_{pc}(k_{ffi} - k_{pv}r_v)}{l_f} i_{o,d} + \frac{\omega_b k_{pc}k_{pv}l_v}{l_f} \omega_g i_{o,q}$$
$$+ \frac{\omega_b k_{AD}}{l_f} \varphi_d + \frac{\omega_b k_{iv}k_{pc}}{l_f} \xi_d - \frac{\omega_b k_{pc}k_{pv}k_q}{l_f} q_m - \omega_b i_{cv,q}\delta\omega_{VSM} + \frac{\omega_b k_{pc}k_{pv}l_v}{l_f} i_{o,q}\delta\omega_{VSM} - \frac{\omega_b c_f k_{pc}}{l_f} v_{o,q}\delta\omega_{VSM} + \frac{\omega_b k_{pc}k_{pv}k_q}{l_f} q^* + \frac{\omega_b k_{pc}k_{pv}}{l_f} \hat{v}^*$$

$$\frac{di_{cv,q}}{dt} = \frac{\omega_b c_f k_{pc}}{l_f} \omega_g v_{o,d} + \frac{\omega_b(k_{ffv} - 1 - k_{AD} - k_{pc}k_{pv})}{l_f} v_{o,q} - \frac{\omega_b(k_{pc} + r_f)}{l_f} i_{cv,q} + \frac{\omega_b k_{ic}}{l_f} \gamma_q - \frac{\omega_b k_{pc}k_{pv}l_v}{l_f} \omega_g i_{o,d} + \frac{\omega_b k_{pc}(k_{ffi} - k_{pv}r_v)}{l_f} i_{o,q}$$
$$+ \frac{\omega_b k_{AD}}{l_f} \varphi_q + \frac{\omega_b k_{iv}k_{pc}}{l_f} \xi_q + \omega_b i_{cv,d}\delta\omega_{VSM} - \frac{\omega_b k_{pc}k_{pv}l_v}{l_f} i_{o,d}\delta\omega_{VSM} + \frac{\omega_b c_f k_{pc}}{l_f} v_{o,d}\delta\omega_{VSM}$$

$$\frac{d\gamma_d}{dt} = -k_{pv}v_{o,d} - c_f\omega_g v_{o,q} - i_{cv,d} + (k_{ffi} - k_{pv}r_v)i_{o,d} + k_{pv}l_v\omega_g i_{o,q} + k_{iv}\xi_d - k_{pv}k_q q_m + k_{pv}l_v i_{o,q}\delta\omega_{VSM} - c_f v_{o,q}\delta\omega_{VSM} + k_{pv}k_q q^* + k_{pv}\hat{v}^*$$

$$\frac{d\gamma_q}{dt} = c_f\omega_g v_{o,d} - k_{pv}v_{o,q} - i_{cv,q} - k_{pv}l_v\omega_g i_{o,d} + (k_{ffi} - k_{pv}r_v)i_{o,q} + k_{iv}\xi_q - k_{pv}l_v i_{o,d}\delta\omega_{VSM} + c_f v_{o,d}\delta\omega_{VSM}$$

$$\frac{di_{o,d}}{dt} = \frac{\omega_b}{l_g} v_{o,d} - \frac{\omega_b r_g}{l_g} i_{o,d} + \omega_b\omega_g i_{o,q} + \frac{\omega_b\hat{v}_g \cos(\delta\theta_{VSM})}{l_g}$$

$$\frac{di_{o,q}}{dt} = \frac{\omega_b}{l_g}v_{o,q} - \omega_b\omega_g i_{o,d} - \frac{\omega_b r_g}{l_g}i_{o,q} + \frac{\omega_b \hat{v}_g \sin(\delta\theta_{VSM})}{l_g}$$

$$\frac{d\varphi_d}{dt} = \omega_{AD}v_{o,d} - \omega_{AD}\varphi_d$$

$$\frac{d\varphi_q}{dt} = \omega_{AD}v_{o,q} - \omega_{AD}\varphi_q$$

$$\frac{dv_{PLL,d}}{dt} = \omega_{LP,PLL}v_{o,d}\cos(\delta\theta_{PLL} - \delta\theta_{VSM}) + \omega_{LP,PLL}v_{o,q}\sin(\delta\theta_{PLL} - \delta\theta_{VSM}) - \omega_{LP,PLL}v_{PLL,d}$$

$$\frac{dv_{PLL,q}}{dt} = -\omega_{LP,PLL}v_{o,d}\sin(\delta\theta_{PLL} - \delta\theta_{VSM}) + \omega_{LP,PLL}v_{o,q}\cos(\delta\theta_{PLL} - \delta\theta_{VSM}) - \omega_{LP,PLL}v_{PLL,q}$$

$$\frac{d\varepsilon_{PLL}}{dt} = \tan^{-1}\left(\frac{v_{PLL,q}}{v_{PLL,d}}\right)$$

$$\frac{d\delta\theta_{VSM}}{dt} = \omega_b\delta\omega_{VSM}$$

$$\frac{d\delta\xi_d}{dt} = -v_{o,d} - r_v i_{o,d} + l_v\omega i_{o,q} - k_q q_m + l_v i_{o,q}\delta\omega_{VSM} + k_q q^* + \hat{v}^*$$

$$\frac{d\delta\xi_q}{dt} = -v_{o,q} - l_v\omega i_{o,d} - r_v i_{o,q} - l_v i_{o,d}\delta\omega_{VSM}$$

$$\frac{dq_m}{dt} = -\omega_f i_{o,q}v_{o,d} + \omega_f i_{o,d}v_{o,q} - \omega_f q_m$$

$$\frac{d\delta\omega_{VSM}}{dt} = -\frac{1}{T_a}i_{o,d}v_{o,d} - \frac{1}{T_a}i_{o,q}v_{o,q} + \frac{k_d k_{p,PLL}}{T_a}\tan^{-1}\left(\frac{v_{PLL,q}}{v_{PLL,d}}\right) + \frac{k_d k_{i,PLL}}{T_a}\varepsilon_{PLL} - \frac{k_d + k_\omega}{T_a}\delta\omega_{VSM} + \frac{1}{T_a}p^* + \frac{k_\omega}{T_a}\omega^* - \frac{k_\omega}{T_a}\omega_g$$

$$\frac{d\delta\theta_{PLL}}{dt} = \omega_b k_{p,PLL}\tan^{-1}\left(\frac{v_{PLL,q}}{v_{PLL,d}}\right) + \omega_b k_{i,PLL}\varepsilon_{PLL}$$

$$(28)$$

The steady state operating point of the system under any combinations of input signals can be found by solving this nonlinear system model with derivative terms set to zero.

Small Signal Model of the Reference VSM

Since the state-space model from (28) is nonlinear, classical stability assessment techniques based on eigenvalues are not directly applicable. Thus, in this section, a corresponding linearized small-signal state-space model is derived in the form given by:

$$\Delta\dot{\mathbf{x}} = \mathbf{A}\cdot\Delta\mathbf{x} + \mathbf{B}\cdot\Delta\mathbf{u}. \tag{29}$$

where the prefix Δ denotes small-signal deviations around the steady-state operating point [30]. The values of the state variables at this linearization point are denoted by subscript '0' when they appear in the matrices. For convenience of notation, the dynamic matrix A is expressed through four sub-matrices according to:

$$\begin{bmatrix} \Delta \dot{\mathbf{x}}_1 \\ \Delta \dot{\mathbf{x}}_2 \end{bmatrix} = \begin{bmatrix} \mathbf{A}_{11} & \mathbf{A}_{12} \\ \mathbf{A}_{21} & \mathbf{A}_{22} \end{bmatrix} \cdot \begin{bmatrix} \Delta \mathbf{x}_1 \\ \Delta \mathbf{x}_2 \end{bmatrix} + \mathbf{B} \cdot \Delta \mathbf{u}$$

(30)

where $\mathbf{A}_{11}$, $\mathbf{A}_{12}$, $\mathbf{A}_{13}$ and $\mathbf{A}_{22}$ are given by (31), (32), (33) and (34), while the $\mathbf{B}$ matrix is given by (35):

$$\mathbf{A}_{11} = \begin{bmatrix} & & & & & & & & & \\ & & & & & & & & & \end{bmatrix}$$

(31)

$$\mathbf{A}_{12} = \begin{bmatrix} 0 & 0 & 0 & 0 & 0 & 0 & 0 & 0 \\ 0 & 0 & 0 & 0 & 0 & 0 & 0 & 0 \\ 0 & 0 & 0 & \frac{\omega_b k_{iv} k_{pc}}{l_f} & 0 & -\frac{\omega_b k_{pc} k_{pv} k_q}{l_f} & \omega_b(-l_f i_{cv,q,0} + k_{pc} k_{pv} l_v i_{o,q,0} - C_f k_{pc} v_{o,q,0}) & 0 \\ 0 & 0 & 0 & \frac{\omega_b k_{iv} k_{pc}}{l_f} & 0 & \frac{\omega_b(l_f i_{cv,d,0} - k_{pc} k_{pv} l_v i_{o,d,0} + C_f k_{pc} v_{o,d,0})}{l_f} & 0 \\ 0 & 0 & 0 & k_{iv} & 0 & -k_{pv} k_q & k_{pv} l_v i_{o,q,0} - C_f v_{o,q,0} & 0 \\ 0 & 0 & 0 & 0 & k_{iv} & 0 & -k_{pv} l_v i_{o,d,0} + C_f v_{o,d,0} & 0 \\ 0 & 0 & 0 & \frac{\omega_b \hat{v}_{g,0} \sin(\delta\theta_{VSM,0})}{l_g} & 0 & 0 & 0 & 0 \\ 0 & 0 & 0 & \frac{\omega_b \hat{v}_{g,0} \cos(\delta\theta_{VSM,0})}{l_g} & 0 & 0 & 0 & 0 \\ 0 & 0 & 0 & 0 & 0 & 0 & 0 & 0 \\ 0 & 0 & 0 & 0 & 0 & 0 & 0 & 0 \end{bmatrix}$$

(32)

$$\mathbf{A}_{21} = \begin{bmatrix} \omega_{LP.PLL} \cos(\delta\theta_{PLL,0} - \delta\theta_{VSM,0}) & \omega_{LP.PLL} \sin(\delta\theta_{PLL,0} - \delta\theta_{VSM,0}) & 0 & 0 & 0 & 0 & 0 & 0 & 0 & 0 \\ -\omega_{LP.PLL} \sin(\delta\theta_{PLL,0} - \delta\theta_{VSM,0}) & \omega_{LP.PLL} \cos(\delta\theta_{PLL,0} - \delta\theta_{VSM,0}) & 0 & 0 & 0 & 0 & 0 & 0 & 0 & 0 \\ 0 & 0 & 0 & 0 & 0 & 0 & 0 & 0 & 0 \\ 0 & 0 & 0 & 0 & 0 & 0 & 0 & 0 & 0 \\ -1 & 0 & 0 & 0 & 0 & -r_v & \omega_g l_v & 0 & 0 \\ 0 & -1 & 0 & 0 & 0 & -\omega_g l_v & -r_v & 0 & 0 \\ -\omega_f i_{o,q,0} & \omega_f i_{o,d,0} & 0 & 0 & 0 & \omega_f v_{o,q,0} & -\omega_f v_{o,d,0} & 0 & 0 \\ \frac{i_{o,d,0}}{T_a} & \frac{i_{o,q,0}}{T_a} & 0 & 0 & 0 & -\frac{v_{o,d,0}}{T_a} & \frac{v_{o,q,0}}{T_a} & 0 & 0 \\ 0 & 0 & 0 & 0 & 0 & 0 & 0 & 0 & 0 \end{bmatrix}$$

(33)

$$\mathbf{A}_{22} = \begin{bmatrix}
-\omega_{LP,PLL} & 0 & 0 & -\omega_{LP,PLL}\begin{pmatrix} v_{o,q,0}\cos(\delta\theta_{PLL,0}-\delta\theta_{VSM,0}) \\ -v_{o,d,0}\sin(\delta\theta_{PLL,0}-\delta\theta_{VSM,0}) \end{pmatrix} & 0 & 0 & 0 & 0 & \omega_{LP,PLL}\begin{pmatrix} v_{o,q,0}\cos(\delta\theta_{PLL,0}-\delta\theta_{VSM,0}) \\ -v_{o,d,0}\sin(\delta\theta_{PLL,0}-\delta\theta_{VSM,0}) \end{pmatrix} \\
0 & -\omega_{LP,PLL} & 0 & \omega_{LP,PLL}\begin{pmatrix} v_{o,d,0}\cos(\delta\theta_{PLL,0}-\delta\theta_{VSM,0}) \\ +v_{o,q,0}\sin(\delta\theta_{PLL,0}-\delta\theta_{VSM,0}) \end{pmatrix} & 0 & 0 & 0 & 0 & -\omega_{LP,PLL}\begin{pmatrix} v_{o,d,0}\cos(\delta\theta_{PLL,0}-\delta\theta_{VSM,0}) \\ +v_{o,q,0}\sin(\delta\theta_{PLL,0}-\delta\theta_{VSM,0}) \end{pmatrix} \\
0 & \dfrac{1}{v_{PLL,d,0}} & 0 & 0 & 0 & 0 & 0 & 0 \\
0 & 0 & 0 & 0 & 0 & 0 & 0 & \omega_b \\
0 & 0 & 0 & 0 & 0 & 0 & -k_q & l_v i_{o,q,0} \\
0 & 0 & 0 & 0 & 0 & 0 & 0 & -l_v i_{o,d,0} \\
0 & 0 & 0 & 0 & 0 & 0 & -\omega_f & 0 \\
0 & \dfrac{k_d k_{p,PLL}}{T_a v_{PLL,d,0}} & \dfrac{k_d k_{i,PLL}}{T_a} & 0 & 0 & 0 & 0 & -\dfrac{k_d+k_\omega}{T_a} & 0 \\
0 & \dfrac{\omega_b k_{p,PLL}}{v_{PLL,d,0}} & \omega_b k_{i,PLL} & 0 & 0 & 0 & 0 & 0 & 0
\end{bmatrix}$$

$$(34)$$

$$\mathbf{B} = \begin{bmatrix}
0 & 0 & 0 & 0 & 0 & \omega_b v_{o,q,0} \\
0 & 0 & 0 & 0 & 0 & -\omega_b v_{o,d,0} \\
0 & \dfrac{\omega_b k_{pc} k_{pv} k_q}{l_f} & 0 & \dfrac{\omega_b k_{pc} k_{pv}}{l_f} & 0 & \dfrac{\omega_b k_{pc}(k_{pv} l_v i_{o,q,0} - c_f v_{o,q,0})}{l_f} \\
0 & 0 & 0 & 0 & 0 & \dfrac{\omega_b k_{pc}(-k_{pv} l_v i_{o,d,0} + c_f v_{o,d,0})}{l_f} \\
0 & k_{pv} k_q & 0 & k_{pv} & 0 & k_{pv} l_v i_{o,q,0} - c_f v_{o,q,0} \\
0 & 0 & 0 & 0 & 0 & -k_{pv} l_v i_{o,d,0} + c_f v_{o,d,0} \\
0 & 0 & -\dfrac{\omega_b \cos(\delta\theta_{VSM,0})}{l_g} & 0 & 0 & \omega_b i_{o,q,0} \\
0 & 0 & \dfrac{\omega_b \sin(\delta\theta_{VSM,0})}{l_g} & 0 & 0 & \omega_b i_{o,d,0} \\
0 & 0 & 0 & 0 & 0 & 0 \\
0 & 0 & 0 & 0 & 0 & 0 \\
0 & 0 & 0 & 0 & 0 & 0 \\
0 & 0 & 0 & 0 & 0 & 0 \\
0 & 0 & 0 & 0 & 0 & 0 \\
0 & k_q & 0 & 1 & 0 & l_v i_{o,q,0} \\
0 & 0 & 0 & 0 & 0 & -l_v i_{o,d,0} \\
0 & 0 & 0 & 0 & 0 & 0 \\
\dfrac{1}{T_a} & 0 & 0 & 0 & \dfrac{k_\omega}{T_a} & \dfrac{k_\omega}{T_a} \\
0 & 0 & 0 & 0 & 0 & 0
\end{bmatrix}$$

$$(35)$$

DYNAMIC OPERATION OF THE REFERENCE VSM IMPLEMENTATION

The functional behaviour of the reference VSM implementation is illustrated in this section by means of numerical simulations of a few relevant cases. A dynamic model of the investigated system and its corresponding small-signal state-space model have been implemented in Matlab Simulink with the set of parameters listed in Table 1. The parameters of the PI current and voltage controller in this table are obtained by the Modulus Optimum (MO) and Symmetrical Optimum (SO) criteria, respectively, according to the approach explained in [14] and [39], assuming a switching frequency of 5 kHz. Similarly, the PI controller parameters of the PLL are calculated by the SO criterion assuming a low pass filter with crossover frequency of 500 rad/s. The inertia time constant of the VSM is selected to emulate a SM with an intermediate inertia representative for a distributed generation unit, while the other parameters are selected by experience-based trial-and-error.

Table 1. Parameters of investigated VSM configuration.

Parameter	Value	Parameter	Value
Rated voltage VS,LL,RMS	690 V	Filter inductance lf	0.08 pu
Rated power Sb	2.75 MVA	Filter resistance rlf	0.003 pu
Rated angular frequency ωb	2π*50 Hz	Filter capacitance cf	0.074 pu
VSM Inertia constant Ta	2 s	Grid inductance lg	0.20 pu
VSM Damping coefficient kd	400	Grid resistance rg	0.01 pu
Current controller gain: kpc, kic	1.27, 14.3	Grid voltage r_{lf}	1.0 pu
Voltage controler gain: kpv, kiv	0.59, 736	Active damping filter ωAD	50 rad/s
Power reference $p*$	0.5 pu	Active damping gain kAD	0.5 pu
Frequency droop gain $k\omega$	20 pu	Virtual Inductance lv	0.2 pu
Reactive power reference $q*$	0.0 pu	Virtual Resistance rv	0.0 pu
Reactive power droop gain kq	0.2 pu	PLL filter $\omega LP,PLL$	500 rad/s
Reactive power filter ωf	1000 rad/s	PLL proportional gain kp,PLL	0.084
Voltage reference $\hat{v}^*$	1.02 pu	PLL integral gain ki,PLL	4.69

Dynamic Response to Change in Loading

In the first simulated case, the dynamic response of the system is examined for a step in the power reference input to the VSM from 0.5 pu to 0.7 pu when the grid frequency and the reference frequency are both equal to 1.0 pu. This case is equivalent to a sudden increase in the input power or torque on the shaft of a SM connected to an infinite bus.

The power reference and the resulting electrical power from the VSM are plotted in Fig. 8, where it is shown that the VSM with the selected parameters exhibits a smooth transient response and reaches steady state conditions in approximately 1 s without any overshoot. The step change in the reference triggers also a dynamic response in the rotating speed of the virtual inertia, as shown in Fig. 9. Indeed, the excess "mechanical" power input is accumulated in the virtual inertia of the VSM, resulting in an increasing speed during the first part of the transient. This leads to an increase in the phase angle between the VSM-oriented reference frame and the grid voltage vector as shown in the upper part of Fig. 10, until the electrical power output from the machine balances its "mechanical" input. When the electrical power reaches the input power and the steady-state power balance of the system is restored, the rotational speed of the virtual inertia returns to the synchronous speed of the grid source, just as for a traditional SM.

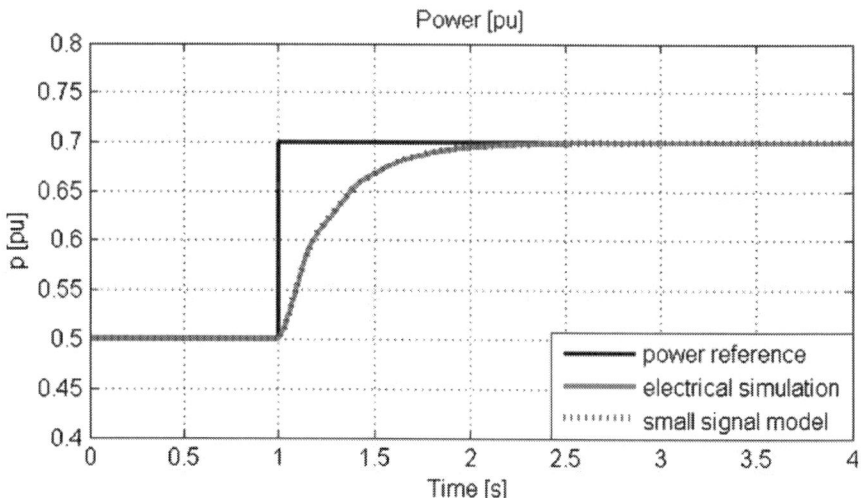

FIGURE. 8. Response of nonlinear model and linearized small-signal model to a step in the power reference input.

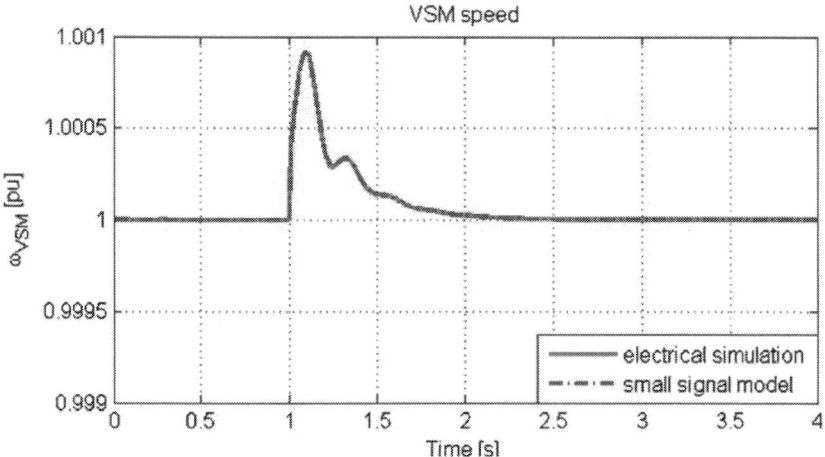

FIGURE. 9. Response of the VSM speed to a step in the power input.

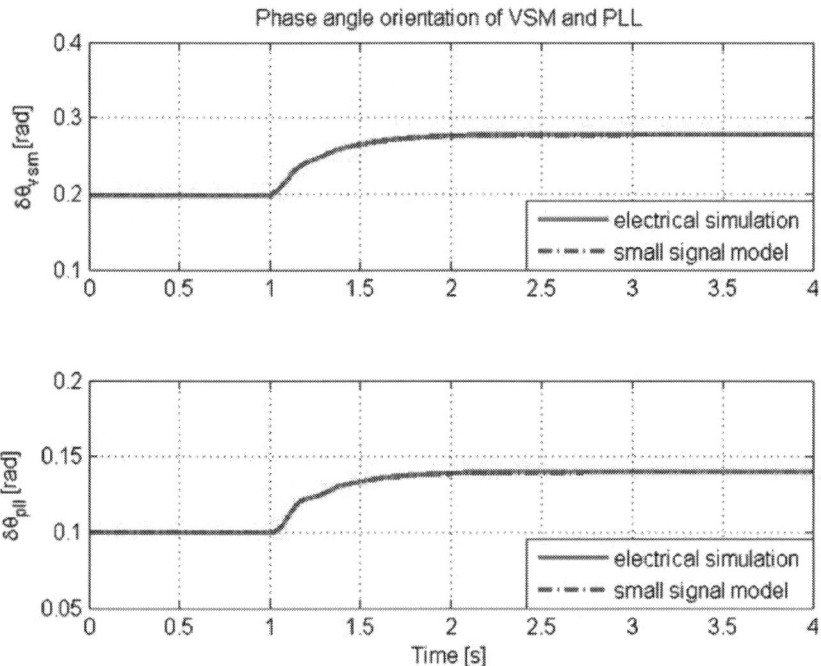

FIGURE. 10. Phase angle displacements of the VSM and PLL reference frames with respect to the grid voltage in response to the step in power reference.

The transient behaviour shown in the presented curves exhibits the same general characteristics as for a conventional SM, but with a more damped response. Indeed, the VSM replicates the behaviour of a classical SM, but its parameters do not have to comply with any physical design constraint. Thus, the parameters of the VSM can be selected with more freedom, without considering any efficiency aspects. In particular, power losses due to the damping effects of the VSM appear only in the control system and not in any physical circuit.

The reference frame orientation of the PLL with respect to the grid voltage vector is shown in the lower part of Fig. 10, illustrating how the phase angle of the voltage at the filter capacitors is changing with the power flow due to the grid impedance. However, it is noticeable how the phase angle of the VSM changes more than for the PLL, due to the virtual inductance included in the VSM. Thus, the VSM phase angle can be considered as equivalent to the phase angle of the internal voltage behind the synchronous reactance of a traditional SM, while the PLL is tracking the phase angle of the voltage at the terminals of the VSM.

The reactive power flow from the VSM is shown in Fig. 11, to illustrate the very small change in the reactive power flow due to the droop controller's response to the change of operating conditions when the active power flow is changed. The plot also indicates small oscillation at a relatively high frequency. This response is mainly due to the response of the LC filter and the measurement filter used in the droop controller.

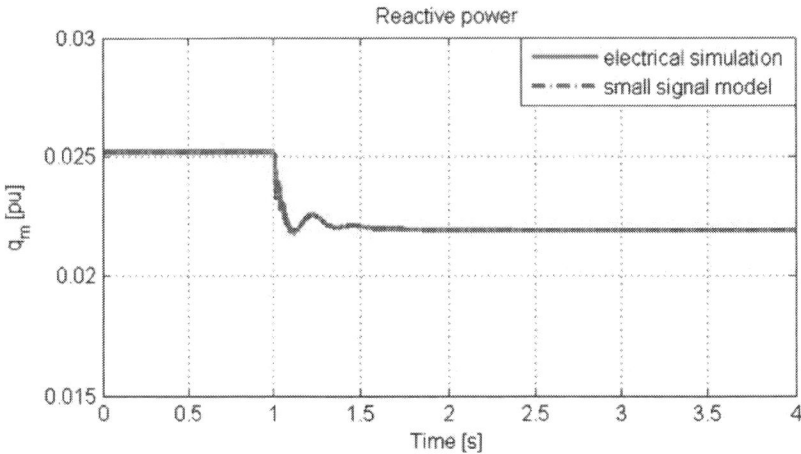

FIGURE. 11. Response of reactive power flow to the step in active power reference.

The presented figures include results from simulation of the non-linear model of the VSM as well as results from simulating the same events with the linearized small-signal state-space model in the same plots. It can be observed that all the curves indicate an excellent match between the two models, for both the fast and slow transients in the system, as long as the operating point stays close enough to the linearization point. This verifies the validity of the linearized model around the steady-state operating point, and indicates that it can be used for investigating system stability and the influence of parametric variations by traditional techniques for linear system analysis.

Response to Change in the Grid Frequency
The results from an additional simulation case are shown in Fig. 12 and Fig. 13, where the grid frequency of the system is ramped down from 1.0 pu to 0.995 pu as shown in the upper part of Fig. 12. Due to this change of grid frequency, the frequency droop gain specified by $k\omega$ is activated and the VSM increases its power output in steady state. The VSM speed follows the grid frequency, releasing energy from its inertia corresponding to the change of speed, and settles at the new operating frequency of the grid with increased power output to contribute to the frequency control of the power system. This can also be seen from the power flow in Fig. 12 and the corresponding converter currents in Fig. 13.

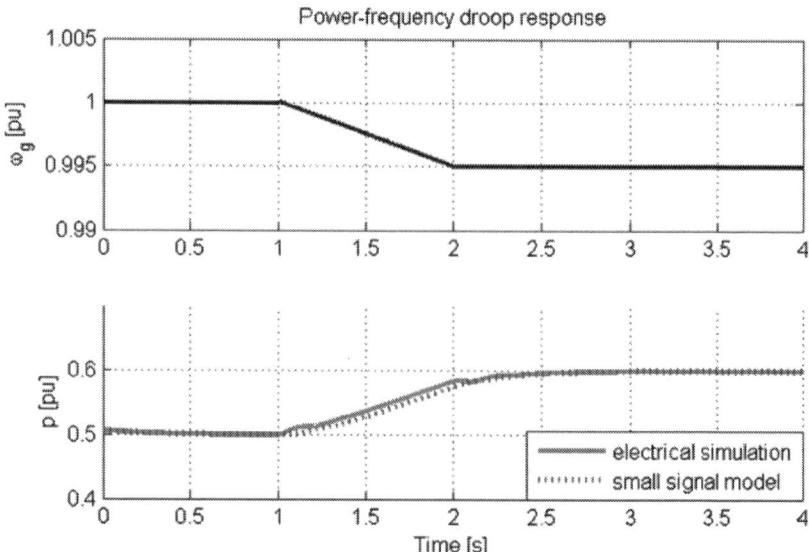

FIGURE. 12. Response of the VSM power flow to a change in the grid frequency.

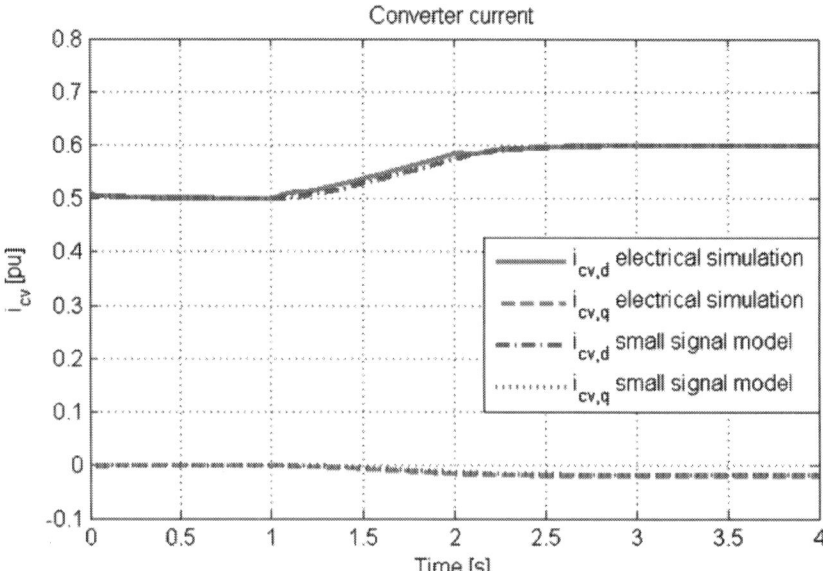

FIGURE. 13. Converter currents as a response to the change of operating conditions when the grid frequency changes.

However, for the linearized small signal model, it should be remembered that the speed deviation from the linearization point in steady state must be zero for the phase angle specified by (5) to settle to a constant value. This implies that the linearized model of the VSM connected to an infinite bus cannot correctly represent the dynamic transition into a new operating point with a different grid frequency than at the linearization point. Thus, a small deviation between the non-linear and linearized models can be observed while the frequency is changing and for a short transient after the frequency has settled to its new steady-state value, as shown by the curves in Fig. 12 and Fig. 13. This is because the grid frequency is not a state variable in the developed small-signal model. Still, the contribution from the change of grid frequency through the B-matrix from (35) is able to represent the corresponding change of power flow also for the linearized model. Thus, the results presented in Fig. 12 and Fig. 13 are demonstrating that the VSM is operating as intended, while at the same time revealing some of the limitations of the developed small-signal model.

System Eigenvalue Analysis
Since the developed linearized small-signal model has been shown to accurately represent the investigated system for small deviations around

the linearization point, the eigenvalues of the **A** matrix can be calculated to systematically identify all the modes of the system. The resulting system eigenvalues for the operating point corresponding to the conditions given in Table 1 are listed in Table 2.

Table 2. System eigenvalues.

$\lambda_1 = -500$	$\lambda_{12} = -224$
$\lambda_{2,3} = -1460 \pm j\,4498$	$\lambda_{13,14} = -6.8 \pm j\,26.4$
$\lambda_{4,5} = -1272 \pm j\,4329$	$\lambda_{15} = -50.8$
$\lambda_{6,7} = -2262 \pm j\,225$	$\lambda_{16} = -50.6$
$\lambda_8 = -1002$	$\lambda_{17} = -37.0$
$\lambda_9 = -470$	$\lambda_{18} = -11.2$
$\lambda_{10,11} = -19.5 \pm j\,245$	$\lambda_{19} = -11.2$

For assessing the system stability, the slow and poorly damped poles are of main interest. It can be noticed from the eigenvalues in Table 2, that the system has several real poles close to the origin and two pairs of complex conjugate poles with small real parts and low oscillation frequency. There are also two pairs of high frequency poles, and one pair of fast poles close to the real axis.

As already seen, the system is stable with the parameters and operating conditions specified in Table 1. Since the small signal model has been shown to accurately represent the system around the linearization point, the system stability for the full operation range of the VSM has been investigated by plotting the system poles when the power reference p^* is swept from -1.0 pu to 1.0 pu. This corresponds to all the steady state operating conditions of the system that can be found by solving the nonlinear model from (28) as a function of the power reference while all other parameter values are as specified in Table 1. The result is shown in Fig. 14, where it can be seen that none of the system poles are moving much when the steady state operating conditions are changed. This indicates that the VSM will exhibit similar dynamic response as already discussed for its entire allowed operating range.

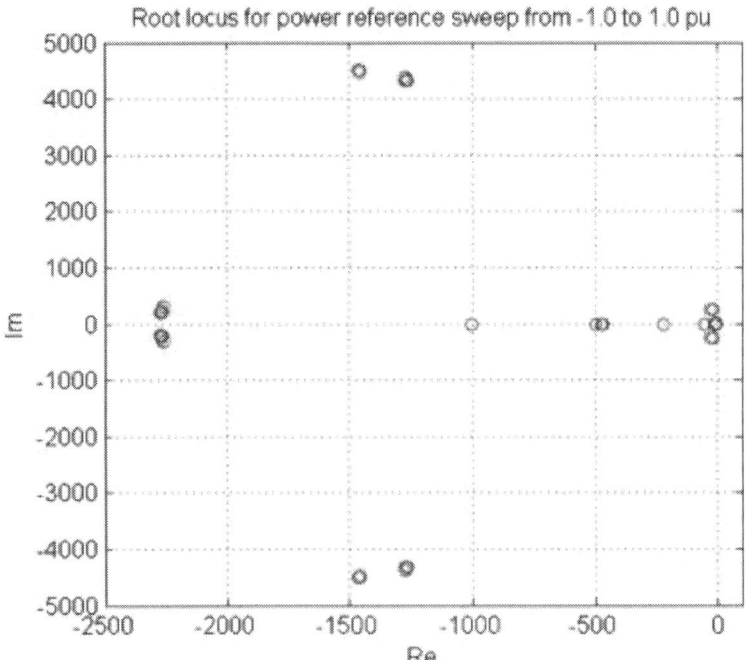

FIGURE. 14. Root locus of all system poles when the power reference $p*$ is swept through the full operating range of the VSM.

Although the system is stable for the full range of operating conditions with the parameters and operating conditions from Table 1, it should be noted that the stability can still be sensitive to variations in the system parameters. For instance, an investigation of a similar configuration reported in [14] indicated that stable operation of the system will be challenging if the switching frequency and thus the bandwidth of the inner loop current controllers is low. In such a case, or due to other changes in the parameters of the inner loop controllers, the poles with the highest imaginary parts, which are associated with the LC filter of the system, can easily move into the right half-plane and cause instability.

There are also other potential parameters that can cause instability in the system under certain conditions. As an example, the trajectory of the poles closest to the imaginary axis are shown in Fig. 15 when the reactive power droop gain kq is swept from 0 to 1.0. The change of the pole location is indicated with a colour gradient, starting from blue when kq is 0 and changing towards red as kq increases. It can be seen from the figure that one pair of complex conjugate poles are moving towards the imaginary axis, and are causing instability when kq becomes too high. Similar investigations of stability limitations can easily be carried out for any of the system parameters, and can be used to support the tuning of the controllers as long as one single or a few clearly identified parameters are responsible for the stability problems in a particular system configuration. Such studies can be of high importance, considering the utilization of the VSM in a SmartGrid scenario with large expected variations in grid configurations, operating conditions and system parameters.

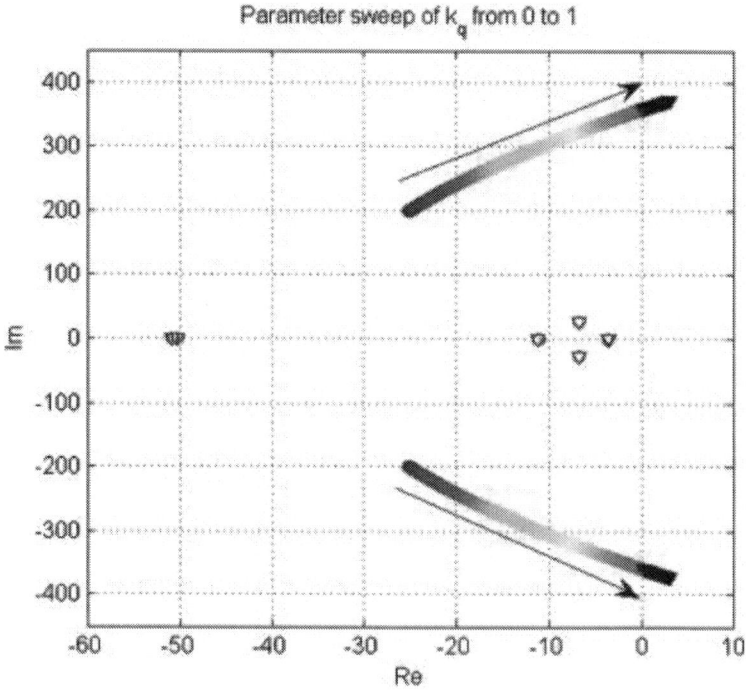

FIGURE. 15. Root locus of the poles closest to the imaginary axis when the reactive power droop gain k_q is swept from 0 to 1.

Parametric Sensitivity of Critical Poles

Investigation of the system stability by ad hoc variations of the controller parameters is challenging for a high order system as the investigated VSM, especially if several parameters are influencing the same pole. To more easily identify the parameters that should be modified to improve the stability and dynamic performance of the system, the parameter sensitivity of pole locations can instead be studied.

The parameter sensitivity of the system poles is defined as the derivative of the eigenvalues with respect to the system parameters. For a dynamic system of order N and with a set of K tunable parameters, the sensitivities define a sensitivity matrix of N by K complex elements. The relative sensitivity $\alpha_{n,k}$ of the parameter ρk with respect to the eigenvalue n can be expressed by (36), where Ψ_n^T and Φ_n are the left and right eigenvectors associated to the eigenvalue λ_n [14] and [30]:

$$\alpha_{n,k} = \frac{\partial \lambda_n}{\partial \rho_k} = \frac{\Phi_n^T \frac{\partial \mathbf{A}}{\partial \rho_k} \Psi_n}{\Phi_n^T \Psi_n}$$

(36)

The real part of these sensitivities is directly associated to the derivatives of the pole location along the real axis with respect to each parameters. Thus, a positive value means that the investigated pole will be moved towards the right by an increase of the corresponding parameter. Similarly, the imaginary part of the sensitivity is associated with the derivative of the pole location along the imaginary axis. However, since the real parts of the pole locations determine the stability and corresponding time constant of the associated system mode, only the real part of the sensitivity matrix is investigated here.

An example of parametric sensitivity analysis for the pole causing instability in Fig. 15 is shown in Fig. 16. From this figure it can be seen that the investigated pole is mainly sensitive to the resistance in the grid and the virtual resistance of the VSM. However, the grid resistance cannot be influenced by the VSM and the virtual resistance should preferably be kept at a low value to achieve a good decoupling of the active and reactive power control of the VSM. Also the actual filter inductance of the converter and the virtual inductance of the VSM have some impact on this pole, but as the grid inductance cannot be influenced by the VSM only the virtual inductance can be modified to improve the stability and dynamic response of the system. The parameter sensitivity also proves that the reactive power droop gain kq has an influence on the pole, and that reducing the droop gain would help to move the pole towards the left. It can also be noticed that an increase of the voltage controller gain kpv can bring the pole towards the left, while all other adjustable parameters have negligible influence on the pole location. Thus, the main parameters that can be used to influence the location of this pole are the droop gain, the voltage controller gain and the virtual inductance.

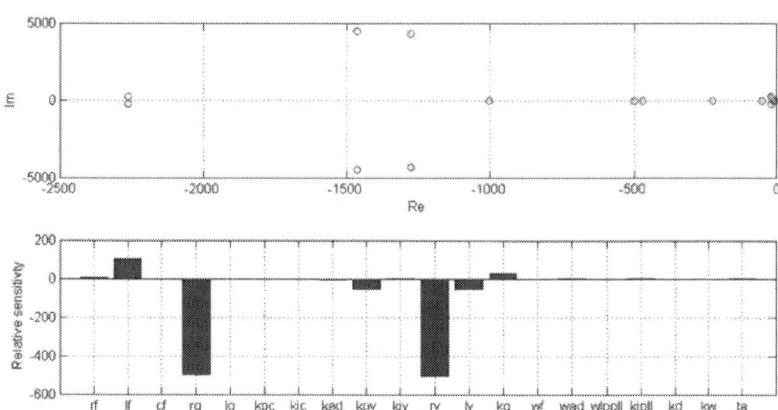

Figure. 16. Parametric sensitivity of the pole resulting in instability with high values of k_q.

Another example of a parameter sensitivity analysis is shown in Fig. 17, for the pole with the highest imaginary part. This is one of the poles that can cause instability due to resonance in the LC filter, even if it is very well damped in this case due to the implemented active damping algorithm. It can, however, be seen that it is sensitive to the physical parameters of the LC filter and the grid, and it is also clear that the location of this pole can be influenced by the proportional gains of the voltage and current controllers, as well as the gain of the active damping algorithm kAD.

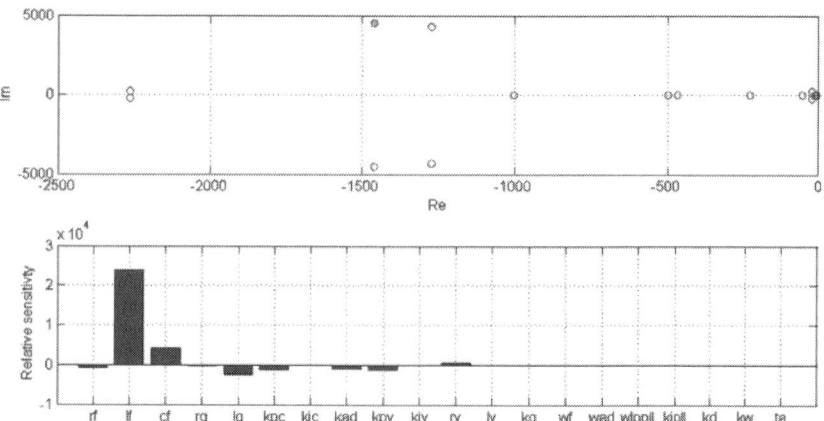

Figure. 17. Parametric sensitivity of the pole with the highest imaginary part.

In case of stability problems in the system, such parameter sensitivity analysis can be easily performed for any of the implicated poles. However, it should be remembered that the change of a parameter to improve the stability conditions or dynamic response associated with one pole, might move another pole in the wrong direction. If that is the case, manual tuning of the controller parameters might be challenging even with support from parameter sensitivity analysis. However, as proposed in [14], it is possible to design automated iterative procedures that can help to bring the pole locations of the system into stable conditions with improved dynamic response. Such approaches might be necessary if a high order system like the investigated VSM is interfaced with other dynamic models and the interaction between the states in the various parts of the overall system model is causing instabilities or poorly damped dynamic responses. The presented VSM model and the tools applied to analyze the properties of the investigated configuration provide a starting point for studying the operation of VSM performance in a larger scale system as will be the case in a SmartGrid scenario.

CONCLUSION

The concept of a Virtual Synchronous Machine (VSM) as an approach for controlling power electronics converters to replicate the behavioural characteristics of Synchronous Machines (SMs) has been introduced during the last decade. This paper has highlighted the inherent advantages of the VSM as a possible alternative for releasing the potential advantages of distributed autonomous control actions of power electronics converters in the SmartGrid context. An implementation of the VSM concept based on an emulated swing equation providing references for cascaded voltage and current controllers has been presented in detail. In particular, a nonlinear mathematical model of the investigated VSM scheme, and its linearized small-signal equivalent, has been derived. These models have been simulated numerically in order to verify and illustrate the behaviour of the VSM and a few of its inherent features. Finally, the system eigenvalues and their parametric sensitivities have been analyzed for the linearized system in order to provide further insight into the dynamic properties and stability characteristics of this VSM implementation. Thus, in addition to the detailed description of a VSM implementation suitable for power system applications in the SmartGrid context, the paper has also presented and verified the models needed for accurately representing the impact of this VSM implementation on the power system in larger scale small-signal power system stability studies.

ACKNOWLEDGEMENT

The work of SINTEF Energy Research in this paper was supported by the project "Releasing the Potential of Virtual Synchronous Machines – ReViSM," through the Blue Sky instrument of SINTEF Energy Research as a Strategic Institute Programme (SIP) funded by the National Basic Funding Scheme of Norway.

REFERENCES

1. CIRED Working Group on Smart Grids, Smart Grids on the Distribution Level – Hype or Vision? CIRED's Point of View, Final Report, 2013, Available at: http://www.cired.at/pdf/CIRED WG SmartGrids FinalReport.pdf

2. European Center for Power Electronics, ECPE, European Power Electronics and Drives Association, EPE, Position Paper on Energy Efficiency – The Role of Power Electronics, 2007, Available at: http://www.ecpe.org/securedl/0/1391290694/35ad20b53952211115c886f006df4d0f595ed7174/ fileadmin/user upload/Public Relations/ECPE Publications/ECPE Position Paper Energy Efficiency.pdf

3. J. Morren, J. Pierik, S.W.H. de Haan, Inertial response of variable speed wind turbines, Electr. Power Syst. Res. 76 (July (11)) (2006) 980–987.

4. H.-P. Beck, R. Hesse, Virtual synchronous machine, in: Proc. of the 9th Int. Conf. on Electrical Power Quality and Utilisation, Barcelona, Spain, 9–11 October, 2007, 6 pp.

5. J. Zhu, C.D. Booth, G.P. Adam, A.J. Roscoe, C.G. Bright, Inertia emulation control strategy for VSC-HVDC transmission systems, IEEE Trans. Power Syst. 28 (May (2)) (2013) 1277–1287.

6. Z. Linn, Y. Miura, T. Ise, Power system stabilization control by HVDC with SMES using virtual synchronous generator, IEEJ J. Ind. Appl. 1 (2) (2012) 102–110.

7. T. Shintai, Y. Miura, T. Ise, Reactive power control for load sharing with virtual synchronous generator control, in: Proc. of the IEEE 7th Int. Power Electronics and Motion Control Conference – ECCE Asia, Harbin, China, 2–5 June 2012, 2012, pp. 846–853.

8. J. Driesen, K. Visscher, Virtual synchronous generators, in: Proc. of the IEEE Power and Energy Society 2008 General Meeting: Conversion and Delivery of Energy in the 21st Century, Pittsburgh, PA, USA, 20–24 July, 2008, 3 pp.

9. Q.-C. Zhong, G. Weiss, Synchronverters: inverters that mimic synchronous generators, IEEE Trans. Ind. Electron. 58 (April (4)) (2011) 1259–1267.

10. S. D'Arco, J.A. Suul, Virtual Synchronous Machines – classification of implementations and analysis of equivalence to droop controllers for microgrids, in: Proc. of IEEE PowerTech Grenoble 2013, Grenoble, France, 16–20 June 2013, 2013, 7 pp.

11. K. Visscher, S.W.H. De Haan, Virtual Synchronous Machines (VSG's) for frequency stabilization in future grids with a significant share of decentralized generation, in: Proc. of the CIRED Seminar 2008: SmartGrids for Distribution, Frankfurt, Germany, 23–24 June 2008, 2008, 4 pp.

12. K. Sakimoto, Y. Miura, T. Ise, Stabilization of a power system with a distributed generator by a virtual synchronous generator function, in: Proc. of the 8th Int. Conference on Power Electronics – ECCE Asia, Jeju, Korea, 30 May–3 June 2011, 2011, 8 pp.

13. H. Bevrani, T. Ise, Y. Miura, Virtual synchronous generators: a survey and new perspectives, Int. J. Electr. Power Energy Syst. 54 (2014) 244–254.

14. S. D'Arco, J.A. Suul, O.B. Fosso, Control system tuning and stability analysis of Virtual Synchronous Machines, in: Proc. of the 2013 IEEE Energy Conversion Congress and Exposition, ECCE 2013, Denver, CO, USA, 15–19 September 2013, 2013, pp. 2664–2671.

15. S. D'Arco, J.A. Suul, O.B. Fosso, Small-signal modelling and parametric sensitivity of a Virtual Synchronous Machine, in: Proc. of the 18th Power Systems Computational Conference, PSCC, Wrocław, Poland, 18–22 August 2014, 2014, 9 pp.

16. F. Blaabjerg, R. Teodorescu, M. Liserre, A.V. Timbus, Overview of control and grid synchronization for distributed power generation systems, IEEE Trans. Ind. Electron. 53 (October (5)) (2006) 1398–1409.

17. J. Rocabert, A. Luna, F. Blaabjerg, P. Rodríguez, Control of power converters in AC microgrids, IEEE Trans. Power Electron. 27 (November (11)) (2012) 4734–4749.

18. D. Pudjianto, C. Ramsay, G. Strbac, Virtual power plant and system integration of distributed energy resources, IET Renew. Power Gener. 1 (March (1)) (2007) 10–16.

19. E.A. Setiawan, Concept and Controllability of Virtual Power Plant (Dr. Ing thesis), Kassel University, Kassel, Germany, 2007.

20. J. Hansen et al., Providing Flexibility with a Virtual Power Plant, Final Demo Report, Deliverable No: 10.3, Twenties Project. Available at:

http://www.twenties-project.eu/system/files/D10.3
Providing%20flexibility%20with%20a%20virtual%20power%20plant.pdf

21. T.L. Vandoorn, J.D.M. De Kooning, B. Meersman, L. Vandevelde, Review of primary control strategies for islanded microgrids with power-electronic interfaces, Renew. Sustain. Energy Rev. 19 (2013) 613–628.

22. T.L. Vandoorn, B. Zwaenpoel, J.D.M. De Kooning, B. Meersman, L. Vandevelde, Smart microgrids and virtual power plants in a hierarchical control structure, in: Proc. of the 2nd IEEE PES Int. Conference and Exhibition on Innovative Smart Grid Technologies, ISGT Europe 2011, Manchester, UK, 5–7 December 2011, 2011, 7 pp.

23. J. Machowski, J.W. Bialek, J.R. Bumby, Power System Dynamics and Stability, Wiley, Chichester, UK, 1997 (Chapters 2 and 5).

24. R. Hesse, D. Turschner, H.-P. Beck, Micro grid stabilization using the Virtual Synchronous Machine (VISMA), in: Proc. of the Int. Conference on Renewable Energies and Power Quality, ICREPQ'09, Valencia, Spain, 15–17 April 2009, 2009, 6 pp.

25. Y. Chen, R. Hesse, D. Turschner, H.-P. Beck, Dynamic properties of the Virtual Synchronous Machine (VSIMA), in: Proc. of the Int. Conference on Renewable Energies and Power Quality, ICREPQ'11, Las Palmas, Spain, 13–15 April 2011, 2011, 5 pp.

26. Y. Chen, R. Hesse, D. Turschner, H.-P. Beck, Investigation of the Virtual Synchronous Machine in the Island Mode, in: Proc. of the 2012 3rd IEEE Innovative Smart Grid Technologies Conference – Europe, ISGT Europe, Berlin, Germany, 15–17 October 2012, 2012, 6 pp.

27. N. Pogaku, M. Prodanovic, ´ T.C. Green, Modeling, analysis and testing of autonomous operation of an inverter-based microgrid, IEEE Trans. Power Electron. 22 (March (2)) (2007) 613–625.

28. S. D'Arco, G. Guidi, J.A. Suul, Embedded limitations and protections for droop-based control schemes with cascaded loops in the synchronous reference frame, in: Proc. of the 2014 Int. Power Electronics Conference, IPEC-Hiroshima 2014 – ECCE Asia, Hiroshima, Japan, 18–21 May 2014, 2014, pp. 1544–1551.

29. S. D'Arco, J.A. Suul, Equivalence of Virtual Synchronous Machines and frequency-droops for converter-based MicroGrids, IEEE Trans. Smart Grid 5 (January (1)) (2014) 394–395.

30. P. Kundur, Power System Stability and Control, McGraw-Hill, New York, 1994.

31. V. Kaura, V. Blasko, Operation of a phase locked loop system under distorted utility conditions, IEEE Trans. Ind. Appl. 33 (January/February (1)) (1997) 58–63.

32. H. Kolstad, Control of an Adjustable Speed Hydro utilizing Field Programmable Devices (Ph.D. thesis), Norwegian University of Science and Technology, 2002.

33. A. Haddadi, G. Joos, Load Sharing of Autonomous Distribution-level Microgrids, in: Proc. 2011 IEEE PES General Meeting: The Electrification of Transportation and the Grid of the Future, Detroit, MI, USA, 24–28 July 2011, 2011, 9 pp.

34. J. He, Y.W. Li, Analysis, design, and implementation of virtual impedance for power electronics interfaced distributed generation, IEEE Trans. Ind. Appl. 47 (November/December (6)) (2011) 2525–2538.

35. V. Blasko, V. Kaura, A new mathematical model and control of a three-phase AC–DC voltage source converter, IEEE Trans. Power Electron. 12 (January (1))(1997) 116–123.

36. M. Malinowski, M.P. Kazmierkowski, S. Bernet, New simple active damping of resonance in three-phase PWM converter with LCL filter, in: Proc. of the 2005 IEEE Int. Conference on Industrial Technology, ICIT 2005, Hong Kong, 14–17 December 2005, 2005, pp. 861–865.

37. N. Kroutikova, C.A. Hernandez-Aramburo, T.C. Green, State-space model of grid-connected inverters under current control mode, IET Electr. Power Appl. 1 (May (3)) (2007) 329–338.

38. S. D'Arco, J.A. Suul, O.B. Fosso, Small-signal modelling and parametric sensitivity of a Virtual Synchronous Machine in islanded operation, IntJ Elec Power (2014), accepted for publication.

39. C. Bajracharya, M. Molinas, J.A. Suul, T.M. Undeland, Understanding of tuning techniques of converter controllers for VSC-HVDC, in: Proc. of Nordic Workshop on Power and Industrial Electronics, NORPIE 2008, Espoo, Finland, 9–11 June, 2008, 2008, 8 pp.

CITATION

Salvatore D'Arco, Jon Are Suul, Olav B. Fosso, A Virtual Synchronous Machine implementation for distributed control of power converters in SmartGrids, Electric Power Systems Research, Volume 122, May 2015, Pages 180-197, ISSN 0378-7796, http://dx.doi.org/10.1016/j.epsr.2015.01.001.

Index